Composite Materials

Production, Properties, Testing and Applications

Composite Materials

Production, Properties, Testing and Applications

K. Srinivasan

Alpha Science International Ltd.
Oxford, U.K.

K. Srinivasan
Department of Metallurgical and Materials Engineering
National Institute of Technology
Karnataka

ALPHA SCIENCE INTERNATIONAL LTD.

7200 The Quorum, Oxford Business Park North
Garsington Road, Oxford OX4 2JZ, U.K.

www.alphasci.com

ISBN 978-1-84265-491-0

Printed in India

**"WORK IS WORSHIP
KNOWLEDGE IS POWER
STRIVE FOR EXCELLENCE"**

Dedicated To My Teachers

PREFACE

Composites are an important class of materials available to humankind. Its study by metallurgists, materials scientists, mechanical, civil & chemical engineers becomes paramount for using them on a large scale in engineering and other applications. The field is so vast that it is truly an interdisciplinary subject in nature and overlaps many areas of metals, ceramics, polymers etc., Composites are simultaneously old & new. Natural composites are as old as creation itself. Modern composite materials had a beginning in the 1930's From then on the progress has been rapid and phenomenal. In recent years its study has assumed a lot of importance.

The present book is an introductory text book for senior undergraduate and junior post graduate students. A prior knowledge of elementary materials is assumed. The first chapter introduces the subject in brief. The second chapter deals with reinforcement. The next five chapters deal with various types of composite materials. The eighth chapter is on mechanics of composites. The ninth chapter elucidates the mechanical properties such as creep, fatigue and fracture toughness. The tenth chapter is devoted to testing of composite materials. The eleventh chapter covers the particulate and dispersion strengthened composites including cermets. Recent developments such as self healing composites are briefly dealt with in twelfth chapter. Numerical problems are considered in the thirteenth chapter. Hints are given for the solution of problems. Some references are given in the end. It may help the reader to dwell deeper in the subject. At the end of each chapter some questions have been appended to help college students who prepare for exams.

The book is to be used by Engineering college students along with class room teaching by experts in the field. It is not intended as a self study manual. For working professionals such as engineers in industries, academicians in colleges and scientists in research organizations this book will be a good first reading to get a basic understanding of the subject. The book touches upon production, properties, testing and applications of composites. By solving the numerical problems one can understand in a limited way the design aspects.

I would like to thank Professor S. Ranganathan of IISc Bangalore, E-G. Ramachandran, K. Srinivasa Raghavan and R. Vasudevan of IIT Madras, Chennai for inculcating in me an ever abiding interest in the area of materials, right from my college days. I am grateful to Professor R. Venugopal of IIT Madras, Chennai for guiding me in the art of preparing manuscripts in engineering subjects. I also acknowledge the indirect contribution of Prof. Kishore who first taught me this subject of composite materials wayback in 1983 at IISc Bangalore. It will not be out of place here if I remember my association with Prof. T.Ramchandran & Prof. N.Venkatraman who were my mentors at National Institute of Technology Karnataka (formerly Karnataka Regional Engineering College), Surathkal. They have been a great source of inspiration to me. My brief stay at University of Arizona, Tucason, U.S.A, as a teaching assistant and interaction with Prof. L.J. Demer and Prof. G.H. Geiger has been fruitful in reinforcing my concepts about metals, materials and manufacturing. I thank Mr. Dinesh Hosadurga of NITK, Surathkal for initial typing and Mr. K.S. Thammaiah of KST Computer Services Surathkal for patiently composing the manuscript. My thanks are due to Mr. B.K Mahesh, Foreman, CAD CAM Lab, NITK, Surathkal for the drawings. Last but not the least is my indebtedness to my wife, Mrs. G. Bhooma and other family members for wholeheartedly supporting and encouraging me during the preparation of the book.

K. Srinivasan

CONTENTS

CHAPTER 1

INTRODUCTION

1.1 DEFINITION

A composite material or a composite is a mixture of two or more distinct constituents all of which are present in reasonable proportions (> 5%) and have different properties so that the composite property is noticeably different from that of each of the constituents. Plastic is not a composite because it is a compound. An alloy is not a composite because it is a homogeneous mixture.

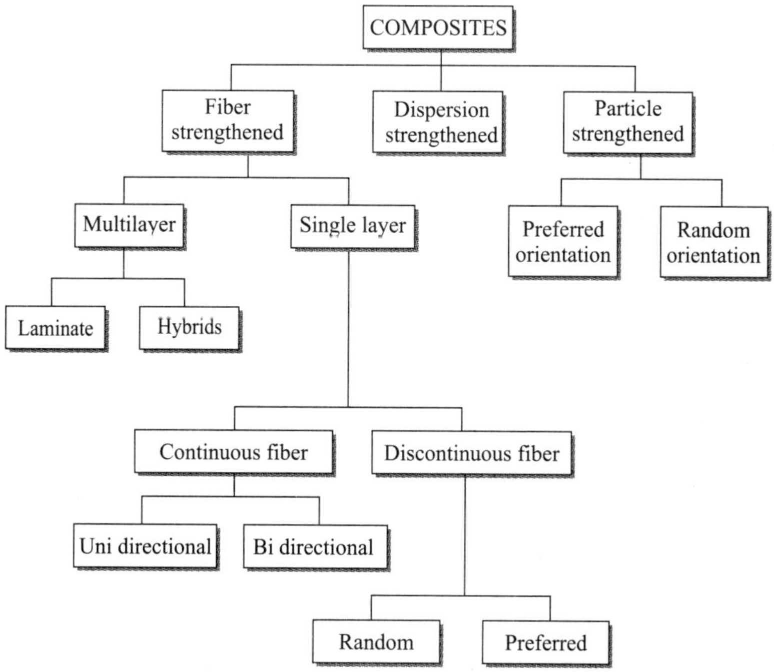

Figure 1.1 Classification of composite according to reinforcements

1.2 CLASSIFICATION OF COMPOSITES

One-way of classifying is based on the nature of the reinforcement employed (Fig 1.1) such as Fibre strengthened (l_f 0.1 – 250 mm, V_f upto 70%) particulate strengthened (d_p > 1.0 mm V_f 5-40%) and dispersion strengthened (d_p = 0.01

– 0.1mm V_f 1–15%). Fibre strengthened composites can be further divided into continuous fibre and short fibre single layer composites, laminates and hybrids (multi-layered composites). Continuous fibre composites can be unidirectional or bidirectional while discontinuous (short) fibre composites can be classified as random orientation and preferred orientation composites. Another way of classification is based on matrix material. Thus there could be ceramic, polymeric, metallic and intermetallic matrix composites. (Fig 1.2)

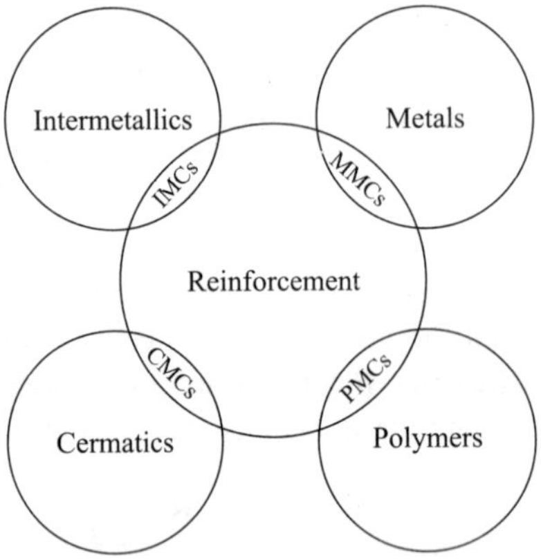

Figure 1.2 Classification of composites according to matrix

1.3 TYPICAL MATRIX AND FIBRE MATERIALS

Metal matrices are reinforced with fibre chiefly to improve their high temperature strength. Polymers are reinforced normally to increase stiffness and strength while ceramic and intermetallic matrices are reinforced mainly to increase the fracture toughness. Common fibres used for reinforcing are B, C, SiC, Al_2O_3, Kevlar, Glass, Polyethylene, Si_3N_4, B_4C, BN, Be, W, Mo etc., The most commonly reinforced metals are Al, Mg and Ti. Some of the ceramics commonly reinforced are LAS,C, Si_3N_4, SiC, Al_2O_3, SiO_2, ZrO_2 and Mullite. The polymers commonly reinforced are (a) thermosets - Epoxy resin and polyester resins. (b) Thermoplastics - Nylon 66, Polypropylene and PEEK. Intermetallic matrix materials, which are usually reinforced, include FeAl, Fe_3Al, Ni_3Al, Ti_3Al, gamma -TiAl, NiAl, and $MoSi_2$.

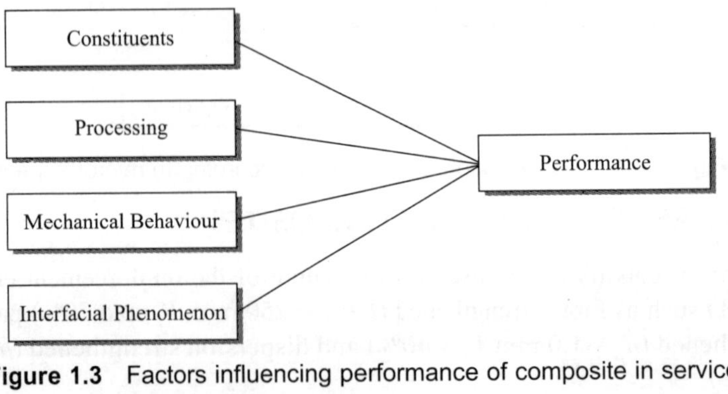

Figure 1.3 Factors influencing performance of composite in service

1.4 APPLICATIONS

Composites are widely used in both military and civilian industries. They are used in the manufacture of a variety of products ranging from aircraft, spacecraft, satellites, missiles and rockets, to sports goods, marine equipments and automobile components. Composites can be designed and fabricated to obtain various properties that suit specific applications. As of now, PMCs are the most widely used composite materials. MMCs come next. CMCs and IMCs are not yet quiet popular but hold good promise for the future.

1.5 LAMINATES AND HYBRIDS

In the context of composite materials, a lamina is defined as a homogeneous orthotropic sheet containing a matrix reinforced by a fibre. A laminated composite is made by stacking a number of such orthotropic sheets in a specific orientation to obtain the desired properties. If each lamina contains a different fibre material with the same matrix or vice versa, it is a hybrid composite.

1.6 COMPARISON OF MMCs, CMCs AND PMCs

Metal matrix composites have many advantages over monolithic metals including a higher specific modulas, higher specific strength, better properties at elevated temperatures, lower coefficient of thermal expansion and better were resistance. But toughness of MMCs is lower compared to metals. MMCs are costly at present. Still MMCs are considered for a wide range of applications in aerospace industries. Mostly MMCs are in development stage or early stages of commercial production. They are not widely established in service like PMCs. Ceramic matrix composites development is not as advanced as that of MMCs or PMCs. The reasons are (1) most of the processing routes for CMCs involve high temperature and therefore we require reinforcements that withstand these high processing temperatures. Therefore ceramic fibers have to be used. No metal fiber can be used since they lose strength and stiffness at intermediate temperatures itself. Toughness of CMCs with ceramic Matrix and ceramic fiber will not be high, even though high temperature strength etc. are not affected and (2) High temperatures required for forming CMCs introduce expansion and contraction thereby creating residual thermal stresses in CMCs which cannot be relived by plastic deformation of matrix unlike MMCs. As a result cracking occurs during processing.

Ideal ceramic matrix materials should be able to

(1) Infiltrate a bundle of fibers, whiskers or particles
(2) Form a mechanical or frictional bond with reinforcement
(3) Have no chemical reaction with the fibre reinforcement during fabrication or service
(4) Not damage the fiber physically
(5) Have good resistance to creep, fatigue and impact

(6) Have high toughness
(7) Be chemically stable – resistant to moisture, hydration, oxidation or volatilization

As on date PMCs are widely developed and used for two reasons. One is that mechanical properties such as strength and stiffness are inadequate for many structural applications in polymers compared to metals and ceramics. By reinforcing polymers considerable improvement in properties can be achieved which is beneficial. Reinforcements also need not have very demanding properties to effect this improvement. The second reason is that processing temperatures and pressures required are low for PMCs. Problems in processing are minimized. Degradation of fibres during manufacture are less and simple equipment are sufficient. PMCs are developed in a record time quickly. In terms of volumes GFRP (Glass Fibre Reinforced Polymers) are second only to concrete in usage among all composites. PMCs are synonymous with the term "composites". Production and usage are phenomenal in quantity.

The performance of a composite depends on several factors. See Fig.1.3

1.7 NOTATIONS

The following notations are frequently used in this book *m*-matrix *c*-composite *f*-fibre, *w*-whisker, *p*-particles. They are used as sub-scripts to indicate that the quantity being referred to pertains to the phase represented by the subscript letter.

CHAPTER 2

REINFORCEMENTS

Reinforcements used in composites are of three types, fibres (continuous & discontinuous) whiskers (single crystal) and particles. These can be metallic, polymeric, ceramic or composite.

2.1 METALLIC FIBRES

Wires of 100 μm diameter can be made with Ti, W, Ta, Mo or steel by conventional wire drawing. Wires of 10 μm diameter or less can be made by Taylor's process. In this process a larger diameter metallic wire is sheathed in glass and treated to a temperature at which the core metal wire melts while the sheath softens. It is then drawn into smaller diameters. Finally the sheath is removed by etching.

The modulus, strength and melting points of metallic materials are important considerations in their use as reinforcement materials.

Metals such as Be, steel, Mo and W show good combination of modulus, strength and refractoriness. Besides, the values are consistent unlike with ceramic materials. However, metallic wires have high density, the exception being Be, which has a high modulus (~ 300 GPa) and low density (1.8 g/cc). However it is expensive and also hazardous due to toxicity. Besides its strength is low (~ 1330 MPa). W wires are characterized by high M.P., high modulus (360 GPa) and very high density (11.3 g/cc). They are also oxidation prone. High carbon (0.9% C) steel wires of very fine diameter (0.1 mm) show a high strength of 5 GPa. Generally, processing metals into fibres is very expensive.

2.2 POLYMERIC FIBRES

Aramids and polythylere fibres are the most commonly used among polymeric fibres.

2.2.1 Aramids

Aramids have the structure of nylon with extra benzene rings in the backbone of the molecular structure, which increase stiffness. The commercial name of aramid is Kevlar. For making fibres, a solution of the aramid in sulphuric acid is held at a low temperature. It is then extruded into a coagulation bath so that the residual solvent is removed. Fibres thus obtained have a low strength of 850 MPa and Young's modulus of 5 GPa. The density would be around 1.45 g/cc. These are stretched and cold drawn to align the structure, which increases the strength to above 2750 Mpa and the modulus to 75 GPa. After this, the fibres can be used upto 300°C. Glass transition temperature Tg is 360°C. Carbonisation occurs at 425°C. The coefficient of thermal expansion as also the electrical and thermal conductivities are low while the thermal capacity is high.

2.2.2 Polyethylene

It has the lowest density among all fibres. It has an elastic modulus of 175 Gpa. Its chemical formula may be represented as follows:

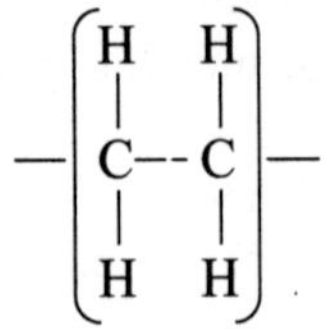

It has a low Meeting Point of 135°C and can be used only up to 100°C.

2.3 CERAMIC FIBRES

Various ceramic materials are used for making fibres.

2.3.1 Glass

Fibres of various varieties of glass such as, E-glass (electrical glass), S-glass (which has high stiffness), C-glass (which is chemical resistant), E-CR-glass (which is electrical and corrosion resistant) and AR-glass (which is alkali resistant) are popular.

For manufacturing glass fibre, the raw materials are melted in a hopper and the molten glass is drawn by gravity through a Pt-Rh bushing containing a large number of holes. The glass fibre that emerges out is given a coating of polymer before winding on a drum. The final glass fibre diameter depends on the orifice diameter of the bushing, viscosity of molten glass and the head level of glass in the hopper. Fibres of 10 μm diameter can be easily produced by the above technique.

2.3.2 Carbon

Carbon fibres can be produced either from polyacrylonitrile (PAN) or from pitch (CN). Although PAN $(CH_2\text{-}CH)_n$ is comparatively expensive, it is more widely used than pitch due to the high yield it offers. Pitch is cheaper and is usually used as the starting material in fibre manufacturing. The steps involved in production are (1) fibreisation (2) stabilisation, (3) carbonisation and (4) graphitisation.

Carbon fibres are given a coating to facilitate easy handling and to improve their adhesion with the matrix material. If uncoated, tiny fragments of C fibre could be air borne and may short circuit electrical machinery.

2.3.3 Alumina Fibre

Alumina fibre can be produced by a sol-gel processing technique. First the Sol concentrate is prepared in the form of a viscous gel. The gel is spun to obtain, the precursor fibre and finally calcined to get the oxide fibre. Following processes have been standardised (1) The 3M process of Minnesota Mining & Manufacturing Company, (2) The Sumitomo Process of Sumitomo chemical company (3) ICI Company process and (4) The Almax Process of Mitsui Mining Company. Some of the other methods available are the Edge defined film fed growth method, Laser - heated floating zone method and the inviscid melt technique. The fibre diameter is usually between 10-12 μm, density of fibre is 3.6 g/cc, UTS is 1.8-2 GPA and elastic modulus is 200-300 GPa.

2.3.4 Silicon Carbide

SiC fibre can be obtained by controlled pyrolysis of a polymeric precursor. This will be very thin compared to what is obtained by Chemical Vapour Deposition (CVD) on W substrate. The fibre thus formed has good mechanical properties, high thermal stability, and oxidation resistance. From Dichlorodimethylsilane, Chlorine is removed by using Na to obtain polydimethyl silane. This is polymerized in an autoclave at 450-500°C, to get polycarbosilane, which is melt spun at 350°C in nitrogen atmosphere to get polycarbosilane fibre. This is cured at 200°C in air or at room temperature in ozone. Then it is pyrolysed at 1300°C in vacuum and cooled to obtain silicon carbide fibre.

2.4 COMPOSITE FIBRES

These are fibres whose compositions are not precisdy homogenous. Fibres made from boron and silicon carbide are examples.

2.4.1 Boron

Composite boron fibres are produced by chemical vapour deposition of boron on a fine, heated tungsten or carbon wire (diameter 12 μm) by reducing boron halide with gaseous hydrogen in a sealed chamber.

$$2BCl_3 + 3H_2 = 2B + 6HCl$$

Boron chloride is very expensive and only about 10% of it is converted into boron in the above reaction. Thus, efficient recovery of the unused halide is a must to make the process economical. Some amount of boron diffuses into the substrate to form borides. The fibre is strictly a composite of boron and the substrate material (W or C). Boron has a density of 2.34 g/cc while boron fibres with a tungsten core possesses a density of 2.6 g/cc for a diameter of 100 μm.

With Laser assisted chemical vapor deposition, fibres of diameter less than 25 μm can be obtained.

2.4.2 Silicon Carbide

By chemical vapour deposition of SiC on a W substrate by reducing alkyl silane (30%) [CH_3SiCl_3] with hydrongen (70%) in mercury sealed chamber one can obtain a SiC fibre. Strictly speaking it is a composite fibre having a core of tungsten.

$$CH_3SiCl_3(g) + H_2\,(g) \longrightarrow SiC(s) + 3HCl(g)$$

To obtain a filament of diameter 100 μm time taken is 20 seconds. The exhaust gases should be properly collected and the unused silane should be recovered for the process to be economical. The diameter of the filament produced could be between 100-150 μm. It is mainly β-SiC with some α-SiC on the W core.

2.5 SiC WHISKER

Whiskers are single crystals which have widely varying morphology and dimensions even when produced by the same process. Whiskers generally have a diameter of 1 μm and length of a few tens of microns. Commercial production of SiC whiskers is done by pyrolysis of rice husks. Volatile constituents are driven off by a coking treatment (heating in an O_2 free atmosphere at 700°C). The raw material is then heated in an inert atmosphere to 1500-1600°C. SiC forms according to the following reaction:

$$3C + SiO_2 \rightarrow SiC + 2CO$$

The residue is heated to 800°C to remove free carbon that contaminates the products. Of the silicon carbide formed, 10% will be in the form of whiskers and the rest will be as particles. The products are separated to obtain 90% whiskers. Whiskers are hazardous to health and must be carefully handled. They are quite costly too, about 5 times more expensive than particles and 100 times more expensive than fibres.

2.6 SiC PARTICLE

Silicon carbide particles are available in plenty. It is also inexpensive. It is widely used in composite materials. It is produced as follows. Silica is a taken in the form of sand and carbon in the form of coke and made to react at a high temperature of 2400°C in an electric furnace. Silicon carbide in this process is in the form of large granules. It is subsequently reduced by comminution to the required size.

CHAPTER 3

METAL MATRIX COMPOSITES (MMCs)

3.1 INTRODUCTION

Metal and alloys are very useful and versatile materials and have been exploited almost fully. Further improvement can be achieved by combining them with reinforcements such as B, C, Al_2O_3 and SiC to produce metal matrix composites. In the sixties, Boron reinforced aluminium was developed and put to use in space shuttle by U.S.A. In the eighties and nineties, fibres such as silicon carbide and aluminium oxide, which are capable of withstanding very high temperatures, were used to reinforce metals at economical prices. The uses of MMCs have been increasing day by day ever since then.

MMCs can be used up to 1000°C, above which only CMCs can be used. MMCs possess the following properties:

(1) Improved temperature resistance compared to pure metals and alloys.
(2) High ductility and strength, good fracture toughness, interlaminar shear strength and transverse tensile strength.
(3) Due to the high elastic modulus of reinforcements very high stiffness compared to monoliths.
(4) Superior electrical and thermal conductivity and also improved magnetic properties.
(5) High dimensional stability due to low thermal expansion coefficient of the matrix.
(6) Resistance to hostile environments such as UV radiation, moisture or chemicals.
(7) Impermeable to gases and liquids.

However, MMCs have some disadvantages such as (I) high density of the metallic matrix (2) complex fabrication processes and (3) high cost of production.

The important matrix materials are (1) aluminium, magnesium, titanium, nickel and super alloys. Some of their properties are given in Table 3.1 below.

Table 3.1 Properties of metals used as matrix

Metal	Density (g/cc)	Melting Point °C	Young's Modulus E (MPa)	Poisson's ratio ν	Thermal conductivity λ (W/mk)
Al	2.7	660	69	0.34	201
Mg	1.8	650	44	0.29	150
Ti	4.6	1660	104-116	0.36	23
Ni	8.9	1453	207	0.36	59
Superalloy (Ni base)	~8.0	~1400	196-214	0.31	8.6-14.9

3.2 PRODUCTION TECHNIQUES

Following are the production routes for MMCs.

(1) Powder Metallurgy (2) Diffusion Bonding	Solid state
(3) Melt Stirring (4) Compo or Rheo casting (5) Squeeze Casting (6) Liquid Infiltration under pressure	Liquid state
(7) Spray Codeposition (8) Insitu Process	

3.2.1 Powder Metallurgy

The matrix and the reinforcement are taken in the form of powders in proper proportion, mixed, heated and subjected to pressure, causing diffusion to take place. The composite so obtained has lower free energy than that of individual matrix and reinforcement powders. This method is well suited for discontinuous fibre, whisker and particulate reinforcements. Theoretically, up to 50% reinforcement can be incorporated in the matrix. However, in practice, the limit is kept at 25% considering the possibilities of damage during processing and loss of toughness in high volume fractions. Al_m/SiC_w is produced by this technique. Refer figure 3.1

3.2.2 Diffusion Bonding

A fibre mat having fibres embedded in a polymer binder is kept between two sheets of matrix foil to form a ply. This can be consolidated or cut into smaller sizes, stacked together in a required sequence and then hot pressed to form the component. It is an expensive process. Only simple shapes such as tubes and plates are made by this technique. Temperatures required are lower than those for liquid state processing techniques such as compocasting or squeeze casting. Al_m/B_f, Ti_m/B_f and Ni_m/B_f are good examples of composites produced by this technique. Refer figure 3.2

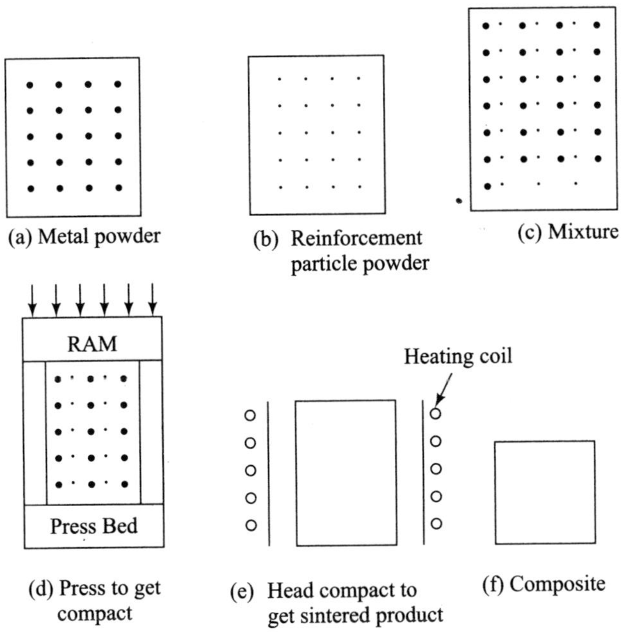

Figure 3.1 Powder metallurgy route to get composite

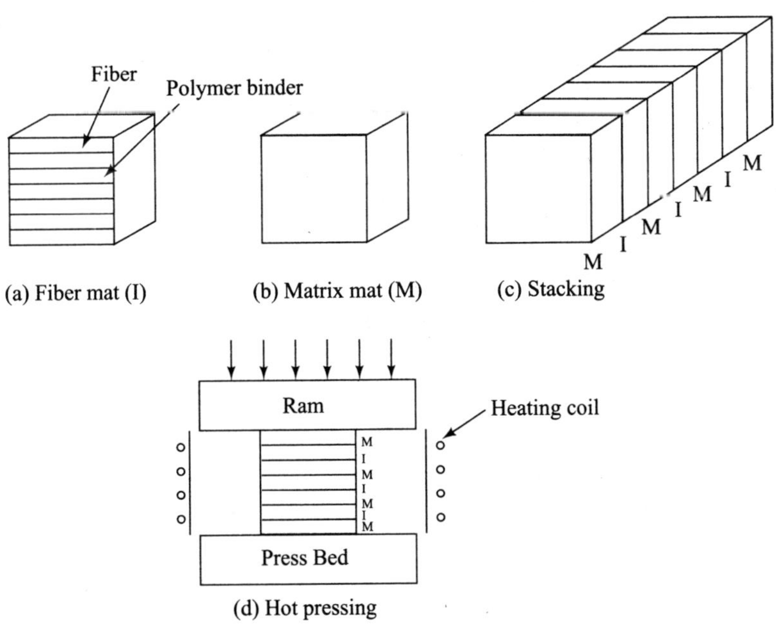

Figure 3.2 Diffusion bonding route to get composite

3.2.3 Melt Stirring

The matrix is taken in molten form to which the reinforcement such as particulates, whiskers or discontinuous fibres are added, thoroughly mixed solidified and then cut into required shapes. This technique involves high temperatures and there is a possibility of damage to reinforcements. Sometimes, molten matrix metal will not wet the reinforcement. Al_2O_3, for instance is not wetted by molten aluminium. Lithium is to be added to increase wettability. Refer figure 3.3

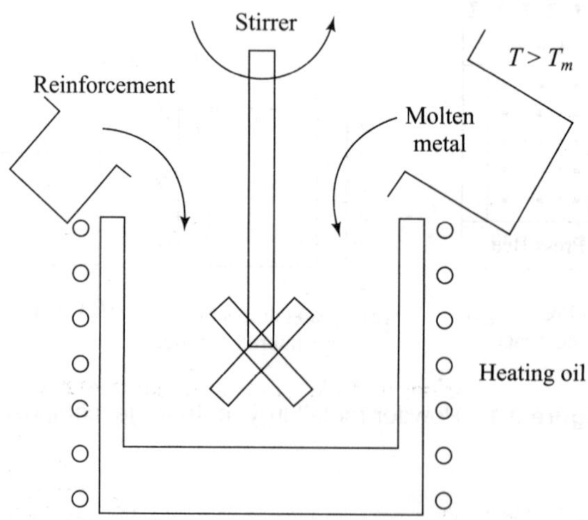

Figure 3.3 Melt stirring route

3.2.4 Compo or Rheocasting

Uniform mixing can be achieved to a greater extent by allowing the melt to cool to a more viscous, two-phase solid-liquid state before stirring. For such a possibility, the liquidus and the solidus should be widely separated as shown in figure below, so that the solidification range over which mushy solid-liquid state exists is very wide. Refer figure 3.4 and 3.5

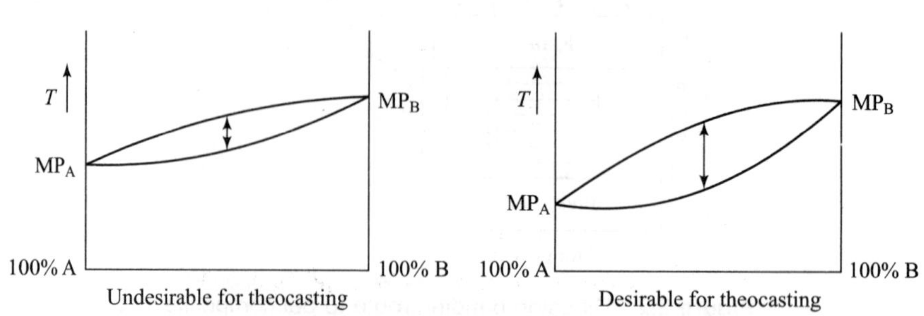

Figure 3.4 Freezing range

The process is based on the fact that it is easy to mix particles more uniformly in a fluid of higher viscosity than in one of lower viscosity. Reinforcement of up to 20% can be incorporated for effective dispersion.

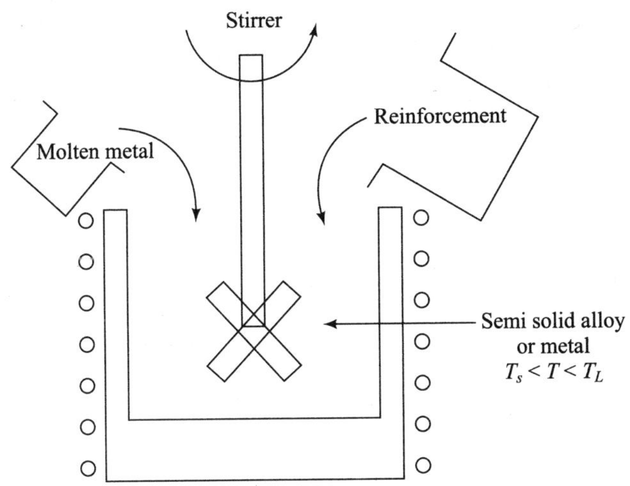

Figure 3.5 Rheo casting route

3.2.5 Squeeze Casting

In this method, a liquid is made to infiltrate a perform by subjecting it to high mechanical pressure. It is suitable for the production of small, net shaped components. Aluminium piston crowns reinforced with discontinuous alumina fibres are produced by this technique. The pressure required for an aluminium alloy matrix is of the order of 70-100 MPa. The process sets a limitation to the size of the components. Refer figure 3.6

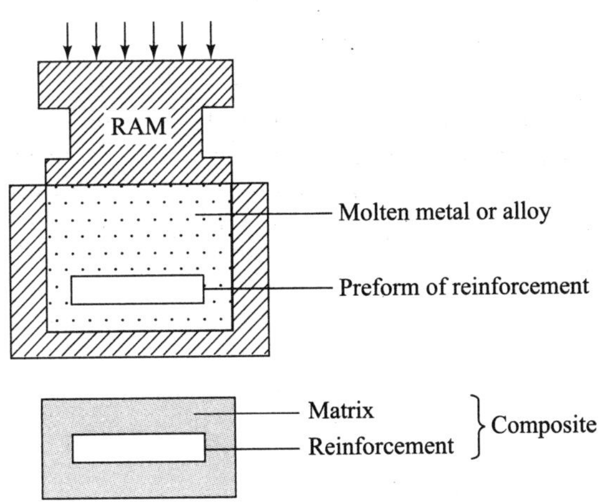

Figure 3.6 Squeeze casting

3.2.6 Liquid Melt Infiltration By Gas Pressure

Here, instead of applying pressure mechanically using a press, a gas is pressurised and forced on the molten matrix metal which then infiltrates the preform containing the reinforcement placed in a hollow die. The process is rapid and no limitation on size exists. Fibre damage by molten metal is reduced due to short contact time and no expensive coatings need be applied to the fibres. Both misalignment and fibre damage are eliminated to a large extent. Refer figure 3.7

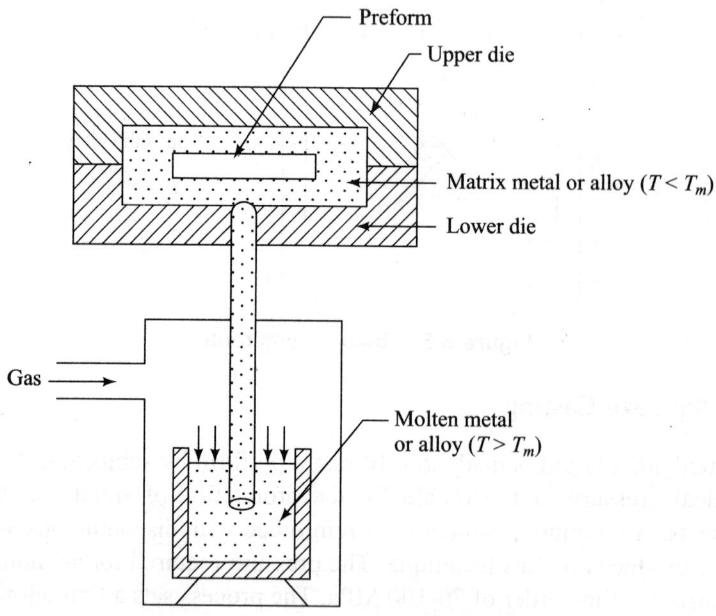

Figure 3.7 Liquid melt infiltration route

3.2.7 Spray Codeposition

Molten matrix metal is atomized and simultaneously sprayed with reinforcement powder on to a substrate. It involves a few milli-seconds and the rapid solidification leads to a matrix with a fine microstructure and reduces the possibility of chemical interaction with reinforcement. High production rate makes the process economical. MMCs with 95% theoretical density can be produced. Refer figure 3.8

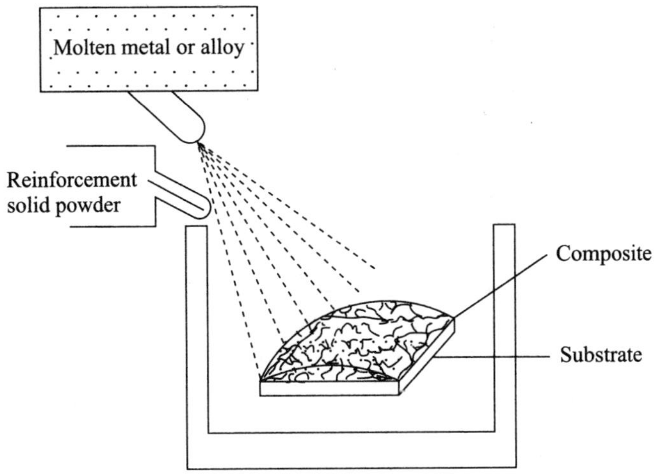

Figure 3.8 Spray co deposition route

3.2.8 In Situ Process

In situ process sometimes effect matrix reinforcement. For instance, unidirectional solidification of eutectic alloys can produce a lamellar or rod like morphology of one phase aligned parallel to the direction of heat removal. However, going strictly by the definition of composites, such products cannot be termed composites. Still they have been included here since their behaviour resembles that of composites with aligned fibres. Refer figure 3.9

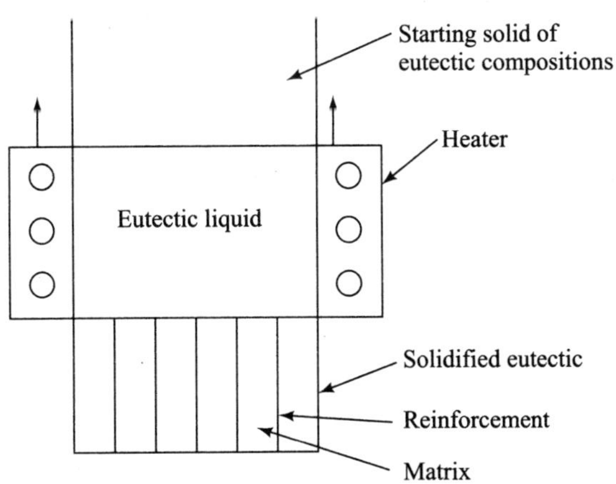

Figure 3.9 Unidirectional insitu solidification route

3.3 PROPERTIES

The coefficient of thermal expansion of metals is usually four to five times larger than that of the reinforcement and therefore composites undergo smaller change in length with temperature than monoliths. Similarly the thermal and electrical conductivities of composites are less than those of monoliths. Thermal deformation resistance, which is defined as the ratio of thermal conductivity to thermal expansion coefficient, is very high for composites than for monoliths. Mg_m / C_f composite is a good example.

MMCs have higher stiffness than monoliths. With increase in percentage of reinforcement, the UTS of a composite increases but the ductility and toughness come down to below those for monoliths. Retention of strength at elevated temperature is to be expected since in manufacturing composites, metals are reinforced with ceramic fibres or whiskers or particles, whose melting points are higher than those of metals. Composites can therefore maintain their strength at high temperatures. Composites also have better creep resistance. Some composites have a typical 3-stage creep curve. Steady state creep stage is limited in some composites. Fatigue strength may be increased or decreased depending on the individual composites.

CHAPTER 4

POLYMER MATRIX COMPOSITES (PMCs)

4.1 INTRODUCTION

Polymer Matrix Composites (PMCs) are more widely used compared to MMCs and CMCs due to the relative ease with which they can be produced and also due to their lightness. The costs are low to moderate. PMCs are widely used in military aircraft, spacecraft (satellites, rockets, space shuttle) and also recently in civilian aircraft. They render space transportation economical by reducing the load to be transported, which in turn, reduces fuel consumption. In helicopters, PMCs are used for rotor blades, rotor heads and fuselage. Future demands are that these composites should be electrically conducting to prevent hazardous damage from lightning and fire resistant to prevent death of air crash survivors from fires that usually follow air crashes.

Generally three classes of polymers namely thermosets, thermoplastics and rubbers are widely used as matrix materials. Weight saving, of about 20% can be achieved by reinforcing these polymers. The resulting composite possess zero thermal expansion apart from lightweight, high strength and high stiffness.

Processing PMCs does not require high pressures or high temperatures as for MMCs or CMCs. Therefore the possibilities of fibre damage and degradation are minimum. Only simple equipments are required to produce PMCs. This has led to the wide popularity of PMCs. By far, glass fibre reinforced polymers are the most widely used composite materials. The main disadvantages of PMCs are, the low service temperature, high sensitivity to radiation and moisture and dimensional instability. However, carbon fibre reinforced polymers have good dimensional stability while epoxies are radiation resistant.

4.2 PRODUCTION OF PMCs

The following methods are widely used for producing PMCs.

(1) Hand Methods
(2) Moulding Methods
(3) Pultrusion Method
(4) Filament Winding Method

4.2.1 Hand Method

There are mainly two methods under this category. The lay up method and the spray up method.

4.2.1.1 Lay up Method

The lay up method requires very little capital equipment and is labour intensive. For limited production this method is very economical. The method is suitable for reinforcement up to 30% by volume. The production process is as follows. Reinforcement in the form of woven rovings or chopped strand mat is placed inside a mould coated with a release agent to minimise sticking. The mould is given a gel coat that provides a decorative and protective surface on the component. A thermosetting resin in liquid form is mixed with a curing agent and applied with a brush or roller in such a way that the matrix gets into the gaps of the reinforcement. Polyesters are widely used as matrix material. Viscosity of the matrix resin and curing time are important from the point of view of production rate, which is characteristically less with this method. Mechanical properties of the products are sometimes adversely affected by trapped air. Reproducing product properties is also difficult. Refer figure 4.1

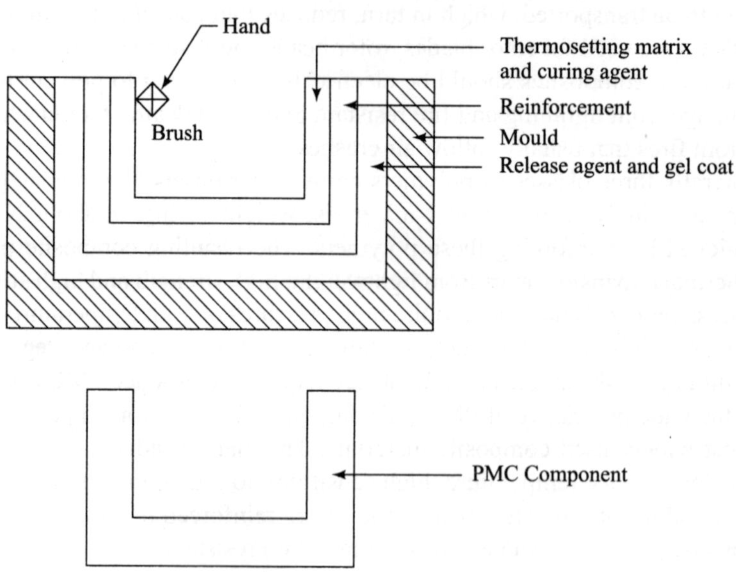

Figure 4.1 Hand lay up method

4.2.1.2 Spray up Method

As against the lay up method, the Spray up method facilitates a higher rate of production. However, with this method, continuous fibres cannot be used. Chopped fibres, resin and catalyst(accelerates the curing) are fed into a gun which sprays the mixture on to the mould which is coated with a release agent for nonsticiking and a gel for a decorative and protective surface. The composite so produced has to be rolled to remove entrapped air and to obtain smooth surface finish. Refer figure 4.2

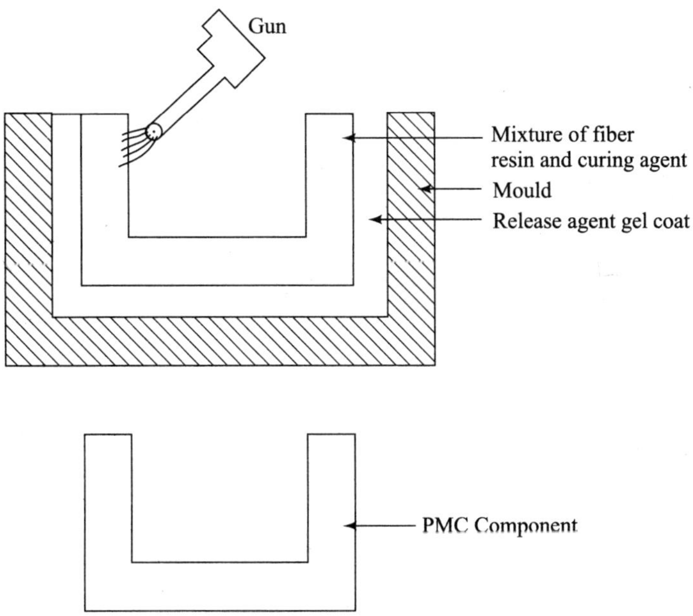

Figure 4.2 Spray up method

4.2.2 Moulding Method

There are a variety of moulding methods. These are expensive but suitable for rapid production.

4.2.2.1 Matched die moulding

A moulding compound is pressed between two matching dies (the male and female counterparts) at suitable temperature to get the required component. There are two types of moulding compounds, namely, Sheet Moulding Compound (SMC) and Dough Moulding Compound (DMC). The former, SMC, is a sheet of resin fibre blend, which contains all the necessary additives such as the curing agent, release agent, and pigment. It is clean to use, gives good product consistency and also survives storage at room temperature for 3-6 months. The latter DMC, is also a blend of all the necessary constituents. It however, uses only short fibres. It has the consistency of a dough or putty and can be fed in required amounts into the dies. The shelf life of DMC is less than that of SMC. Refer figure 4.3

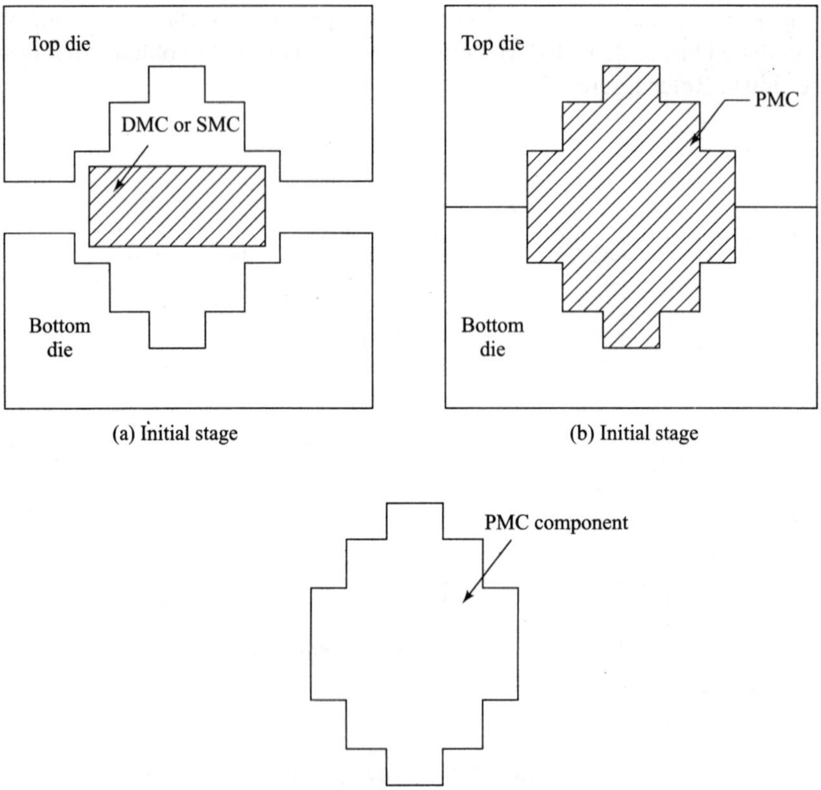

(a) Initial stage (b) Initial stage

(c) Final component

Figure 4.3 Matched die moulding

4.2.2.2 Bag Moulding Methods

In this technique pressure is utilised for making PMC components. These include techniques such as pressure bagging Refer figure 4.4a vacuum forming Refer figure 4.4b and autoclave moulding Refer figure 4.4c. In pressure bagging, though pressure greater than atmospheric pressure is used, no autoclave is used. The process is quite economical. A flexible bag is placed over the lay up on the mould. Gas is pumped into the bag, which causes compression of the lay up into the mould. In vacuum forming, a single die is used and atmospheric

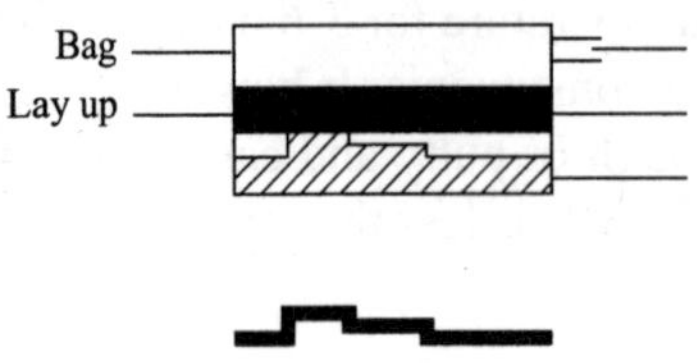

Figure 4.4 (a) Pressure bagging

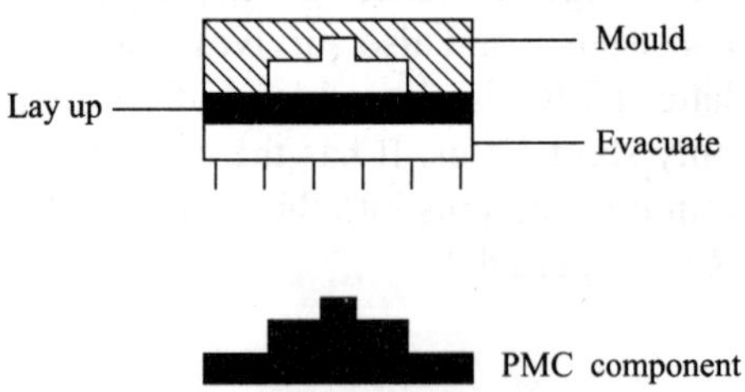

Figure 4.4 (b) Vacuum forming

pressure is employed. A heated preimpregnated reinforcement is sealed on to a mould by a flexible membrane and air is removed, from the die cavity. The impregnated reinforcement is pressed on to the die and vacuum is maintained till it is cured.

In autoclave moulding, pressure greater than atmospheric pressure is used. High density and reproducible products can be manufactured by this technique. The autoclave offers facilities for heating and pressurizing by gas up to 5 MPa. The mould is kept inside the autoclave.

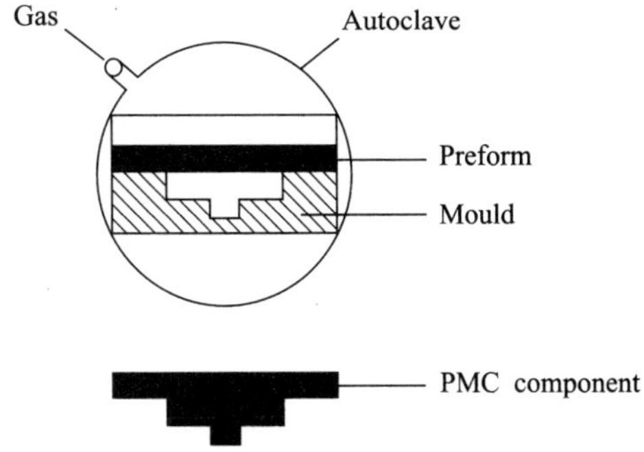

Figure 4.4 (c) Autoclave moulding

4.2.2.3 Closed Mould Methods

The reinforcement is placed inside a closed mould and the matrix material infiltrates it. Three types of techniques can be employed; resin transfer moulding, (Refer figure 4.5) vacuum assisted resin injection moulding (Refer figure 4.6), reinforced reaction injection moulding (Refer figure 4.7.) In resin transfer moulding (RTM) a low viscosity resin is injected into the Glass Reinforced Plastic (GRP) mould and then cured. Complex shapes and large size components can be produced. However, the cycle time is long and may run into days. Only low pressures are employed in RTM. Vacuum assisted resin injection is a variation of this method. The third method, reinforced reaction injection moulding requires shorter cycle time as curing is faster, usually, 1-2 minutes. A preform is kept inside a mould and the matrix resin (obtained by mixing two fast reacting constituents in a high pressure chamber) is injected into the mould at 1 MPa pressure followed by curing. This technique can be employed as a short run process to produce simple components or as a long run process to produce complex components.

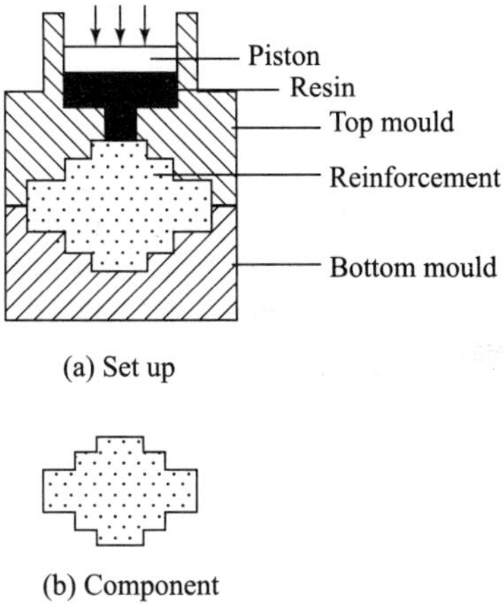

Figure 4.5 Resin transfer moulding

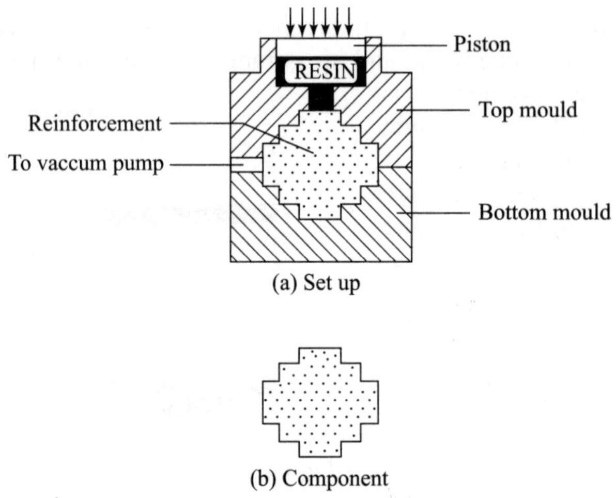

Figure 4.6 Vacuum assisted resin transfer moulding

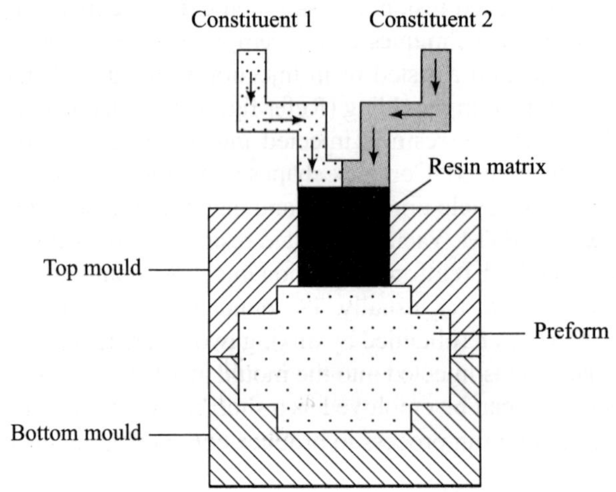

Figure 4.7 Reinforced reaction injection moulding

4.2.3. Pultrusion

This method can be used to produce long rods of uniform cross section. Continuous rovings of the reinforcement are passed through a resin matrix bath to obtain, impregnated reinforcements, which are passed through a heated die that compacts and shapes them to the required profile. The process is somewhat similar to extrusion. Curing takes place in the die and is completed in an oven. Pultrusion is a continuous process by which several meters of rod length can be produced in a few minutes. Refer figure 4.8

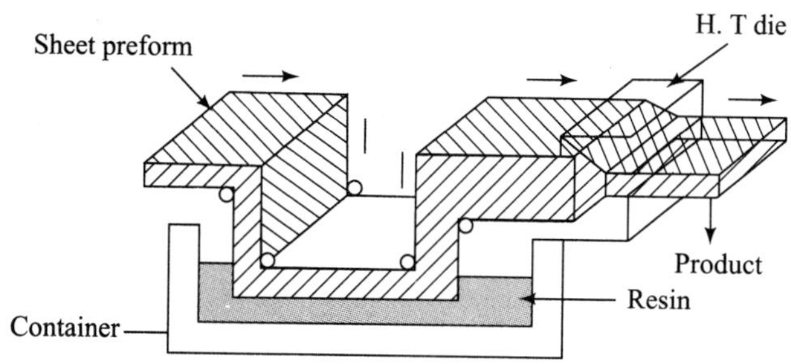

Figure 4.8 Pultrusion of PMC sheet

4.2.4 Filament Winding

A continuous strand of impregnated fibre or tape is wound on to a mandrel and the resin is cured. Subsequently the mandrel is withdrawn. Components such as pipe work, pressure vessels, storage tanks, and aerospace parts such as helicopter blades can be produced by this method. In wet winding, fibres are impregnated as in putrusion. In dry winding preimpregnated reinforcements are used. Wet winding is highly flexible and economical but is not a clean process like dry winding. Refer figure 4.9

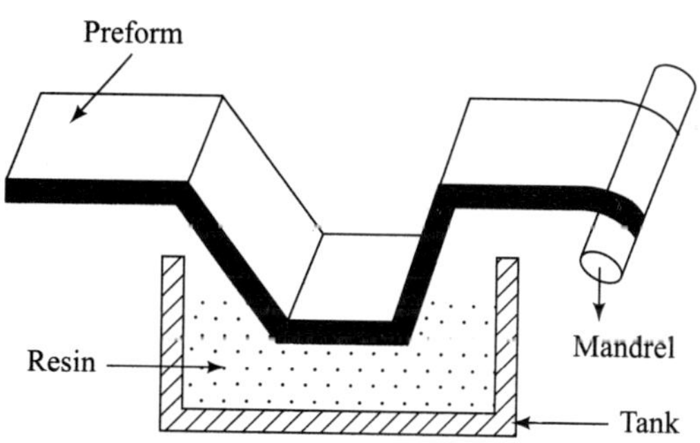

Figure 4.9 Filament winding

CHAPTER 5

CERAMIC MATRIX COMPOSITES (CMCs)

5.1 INTRODUCTION

Ceramics as a class of materials include a wide variety of inorganic materials. They are in general non-metallic and are processed very often at high temperatures. The term ceramics refers to the traditional ceramic materials such as bricks, pottery and tiles as also to the recent additions such as the oxides, nitrides and carbides of silicon, aluminium, zirconium etc. to manufacture which, high quality raw materials and sophisticated chemical processing methods are called for. The latter categories of ceramics have high resistance to heat, chemicals, and abrasion and wear which makes it difficult to fabricate them economically and in a simple way. These ceramics, sometimes called 'advanced ceramics', are mainly used as matrix materials to manufacture ceramic matrix composites. High strength and hardness, excellent high-temperature properties, chemical inertness, wears resistance and low density makes them suitable for the purpose. However certain drawbacks do limit their use as ceramic matrix materials. They have poor ductility and plasticity, cannot withstand tensile and impact loading and are prone to catastrophic failure under thermal or mechanical shock loading. This behaviour is generally termed as 'not forgiving'. This is due to the absence of any energy-dissipating phenomenon in ceramics. To impart damage tolerant behaviour to ceramics was the major concern of the materials community during the last two decades and this led to the development of CMCs with improved properties. Thermosetting and themoplastic materials can be used upto 300°C, while metals and alloys can be used upto 900°C. As against this, ceramics can be used upto 1500°C. Clearly ceramics are the only materials that can be used at such very high temperatures. Ceramics are generally characterized by high hardness, strength and elastic modulus low density, low thermal expansion coefficient as well as low thermal and electrical conductivity. Ceramics are inherently brittle as evident from their low K_{IC} values. See Table 5.1 below for the properties of metals, ceramics and polymers.

Toughness of ceramics needs to be increased for almost all applications and an abvious solution is to go for CMCs.

Table 5.1 Comparison of Properties of Metals, Ceramics and Polymers

	Young's Modulus (Gpa)	Poisson's Ratio ν	Thermal Conductivity λ (W/mK)	Thermal Expansion Coefficient $\alpha\ 10^{-6}K^{-1}$	Densityγ ρ Mg/m^3	Fracture Toughness K_{IC} $MPam^{1/2}$
SiC	480	0.3	-	4	3.2	2-4
Al_2O_3	380	0.25	29	8	3.9	1-3
Si_3N_4	310	3.1	3.1	2.5-3.5		
Low C steel	210	0.29	63	15.0	7.86	50
Al alloys	71	0.34	201	23.0	2.71	20-50
Ti alloys	104-116	0.36	23	9.0	4.54	50-100
Cast iron(grey)	110	0.27	75	11.0	7.15	4-10
Polyetheylene:						
LD	0.18	-	-	250.0	0.920	1-2
HD	0.43	-	-	250.0	0.955	
Nylon	4.5		0.25	-	1.15	3
Epoxy	2.1-6.0	-	-	39.0	1.15	0.5
Polyester	1.3-4.5	-	-	150	1.1-1.5	0.5

Common ceramic matrix materials include the various glasses, glass-ceramics and ceramics such as carbon, silicon carbide, silicon nitride, aluminides and oxides. Reinforcements used are carbon, silicon carbide, titanium diboride, silicon nitride and alumina. Incorporation of fibres, whiskers and particles in a ceramic matrix results in a tougher material due to the development of energy dissipating mechanisms such as fibre bridging, crack deflection, fibre pull out etc. The ratio of elastic modulus of matrix to reinforcement is low for ceramic matrices as against metal or polymer matrices. It is often equal to unity or less than unity in case of CMCs. It may be noted that the load bearing capacity of the reinforcement and that of the matrix are almost the same in the case of CMCs. However, the main objective in manufacturing CMCs is to increase the toughness, not strength.

5.2 PRODUCTION TECHNIQUES

The following production routes are available for CMCs. The conventional techniques are

(1) Cold pressing and sintering
(2) Hot pressing
(3) Reaction bonding
(4) Combination of (2) and (3)
(5) Hot pressing for whisker reinforced composites.

The new techniques are:

(1) Infiltration techniques such as Liquid infiltration, Squeeze casting and Pultrusion,
(2) Lanxide process, (3) In situ, chemical process such as Chemical Vapour Impregnation (CVI) and Liquid phase sintering, (4) Sol Gel techniques, (5) Polymer pyrolysis, (6) Self propagating High temperature Synthesis (SHS).

5.2.1 Conventional Processes

There are a number of conventional methods for the production of CMCs. Some of these are described herein.

5.2.1.1 Cold pressing and sintering

Cold pressing of the matrix powder (ceramic powder) and reinforcement powder is first done, followed by the application of temperature (sintering). The ceramic powder is mixed with an organic binder. The binder must be removed completely to achieve a fully sintered component with near theoretical density. Refer figure 5.1. Matrix is taken in powder form. Reinforcement is taken as particles or whiskers. Both are thoroughly mixed with a binder and put in a cylindrical die. Using a punch this mixture is compacted is a press. Single action pressing or double action pressing can be done. Later gives better distribution of reinforcement and uniform density. This compact is sintered in a furnace at a temperature $T \geq 0.8\ T_m$ and less than T_m. Shrinkage occurs. Porosity is eliminated. Density is increased to near about theoretical density.

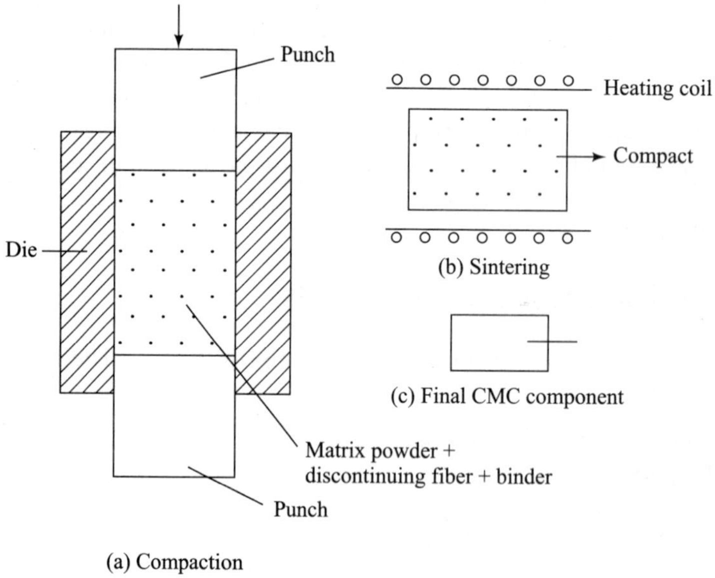

Figure 5.1 Cold pressing

5.2.1.2 Hot pressing

It is also called slurry infiltration technique. It involves two stages; incorporation of a reinforcing phase into an unconsolidated matrix and consolidation by hot pressing. The process is well suited for glass and glass-ceramic matrix composites due to the low temperature involved. The process involves the following steps.

Fibre preform is impregnated with a matrix containing slurry by passing it through a slurry tank. The slurry consists of the matrix powder, a carrier liquid (water or alcohol) and an organic binder. The impregnated preform (called prepregs) is wound on a drum and dried. It is then cut and stacked and the binder is burnt out at 500°C. It is then consolidated by applying pressure at 800-925°C. A fibre / glass composite is obtained. Refer figure 5.2

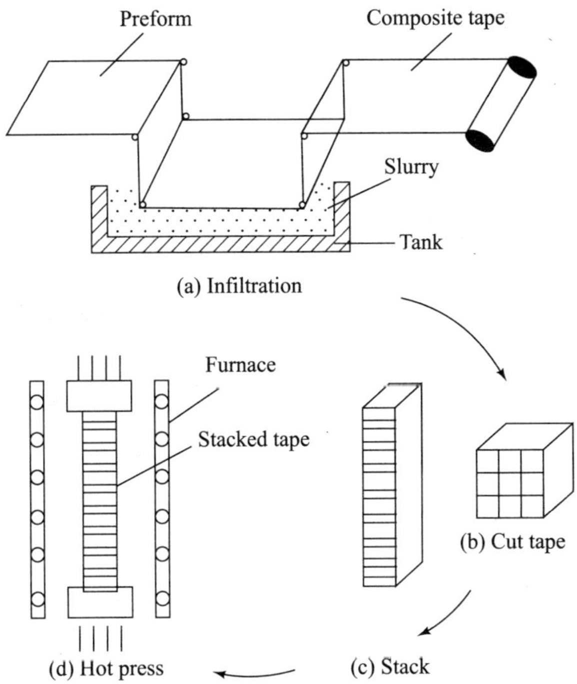

Figure 5.2 Slurry infiltration and hot pressing

5.2.1.3 Reaction bonding

A loose compact of solid powder is reacted with a liquid or gas to obtain the desired CMC. No shrinkage of matrix occurs during densification. This technique is mainly used for SiC or Si_3N_4 matrix composites. It has the advantage that a large volume fraction of whiskers or fibres can be used. Multi-directional continuous fibre preforms can be used. Fibre degradation is minimised since the temperature required for the bonding reaction is less than the sintering temperature. The biggest drawback is the high porosity of the products which is difficult to avoid. Refer figure 5.3 Silicon is taken in powder form and mixed with SiC_w & compacted to desired shape. It is heated and nitrogen gas is passed in a furnace at a temperature of 1100^0-1400^0C. Following reactions take place

$$3Si(s) + 2N_2(g) = Si_3N_4(s)$$
$$3Si(g) + 2N_2(g) = Si_3N_4(s)$$
$$Si(s) + SiO_2 = 2SiO(g)$$

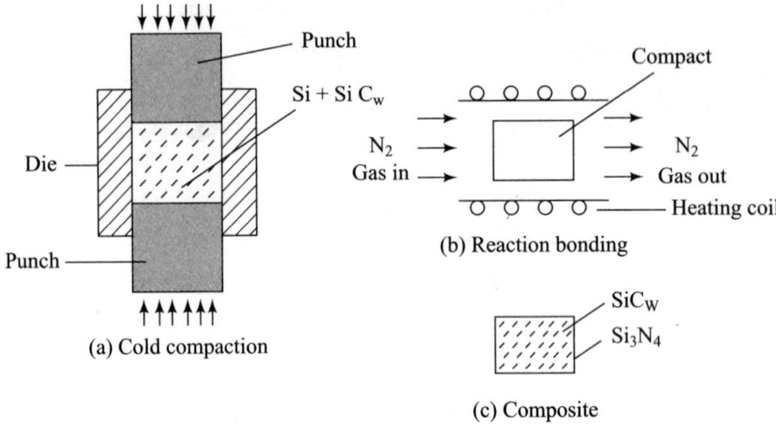

Figure 5.3 Reaction bonding

The last reaction accounts for silicon loss. Thus silicon nitride matrix is formed and we get a composite of silicon nitride and silicon carbide.

5.2.1.4 Combination of (2) and (3)

Another technique for producing CMCs is a hybrid process involving a combination of hot pressing and the reaction bonding technique. Silicon cloth is prepared using a mixture of silicon powder, a polymer binder and an organic solvent. Fibre mats are obtained by filament winding of SiC with a binder. The SiC fibre mats and silicon cloth are stacked and the binder is removed. Then it is hot pressed in a die in vacuum or in N_2 atmosphere. The silicon matrix is converted to Si_3N_4 by nitriding at 1100-1400°C. The matrix has about 25% porosity. Refer figure 5.4

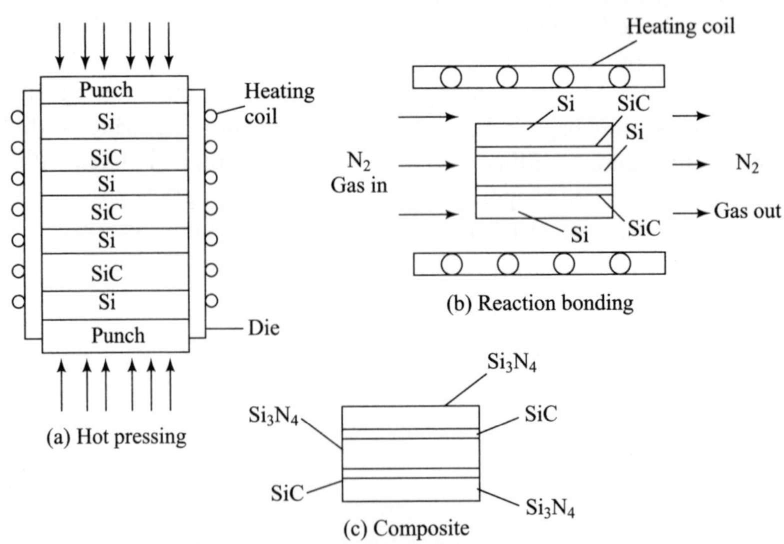

Figure 5.4 Combination of hot pressing with reaction bonding

5.2.1.5 Hot pressing for whisker reinforced composites

Whisker is mixed with ceramic powder slurry. This is followed by drying and hot pressing. SiC_w / Si_3N_{4m} composite is produced by this technique. The process involves HIPping as the last step. Refer figure 5.5. The ceramic matrix is taken in powder form, thoroughly mixed with whisker and hot pressed by simultaneous application of heat and pressure. Compaction and sintering takes place at the same time. Hot Isostatic pressing is also done afterwards. The pressure will be uniform on all sides. Better distribution and uniform density can be achieved. Alternatively whisker is mixed with ceramic powder, made into a slurry and dried. It is then hot pressed by applying pressure and temperature simultaneously followed by hot isostatic pressing.

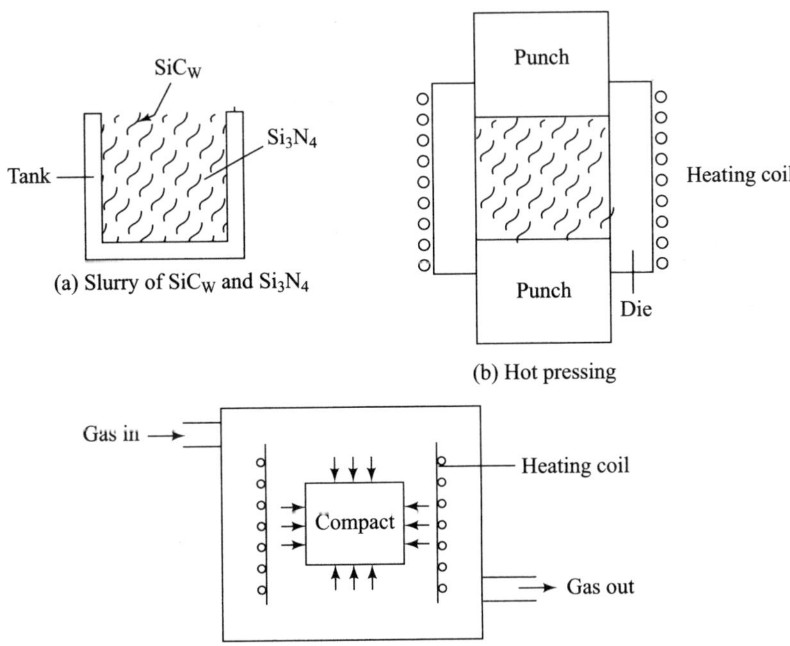

Figure.5.5 Hot pressing for whisker reinforced composites

5.2.2 New Techniques

Following is a brief account of the various new techniques for the production of CMCs.

5.2.2.1 Infiltration

There are various methods, which employ infiltration for producing composites.

(a) Liquid infiltration: A fibre preform is infiltrated by a liquid ceramic matrix material to produce the composite. Porosity is almost completely absent and a high-density matrix is obtained. Temperatures involved in this process are high, since melting of ceramic is required. The high viscosity of molten ceramic is the major factor that renders infiltration difficult. It must be ensured that the liquid ceramic matrix properly wets the reinforcement. Refer figure 5.6. The widely used technique under liquid infiltration methods is Matrix Transfer Moulding. Sheets and tubes can be produced. A preform made of silicon carbide cloth is wound on a mandrel and inserted from the bottom into a cylindrical mould while a glass slug is inserted into the mould from the top. Heat is applied and the glass is molten. Pressure is applied to force the fluid glass into the porous preform. After cooling the composite cube is ejected from the mould. Refer figure 5.7. Similarly a composite sheet also can be produced.

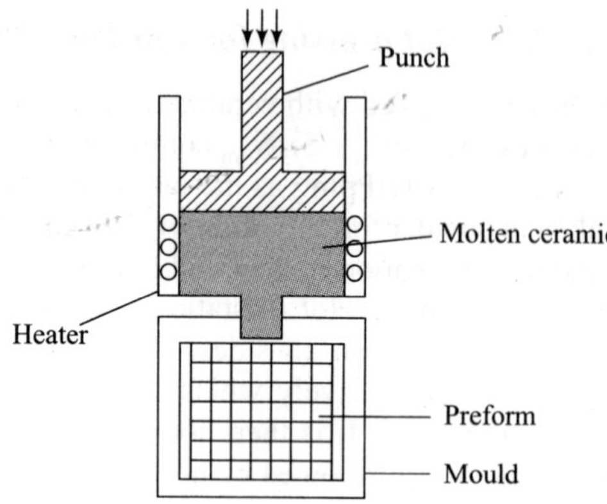

Figure 5.6 Liquid melt infiltration

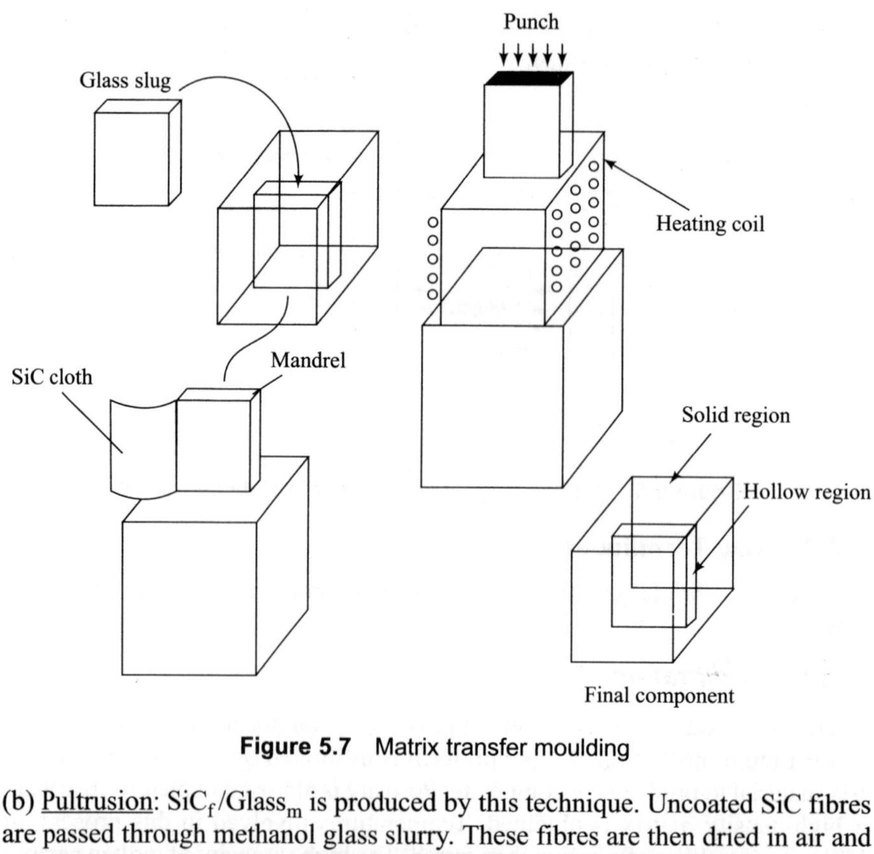

Figure 5.7 Matrix transfer moulding

(b) Pultrusion: $SiC_f/Glass_m$ is produced by this technique. Uncoated SiC fibres are passed through methanol glass slurry. These fibres are then dried in air and pulled through a molten glass bath to obtain a continuous $SiC_f/Glass_m$ composite rod. Advantages of this process are that a homogeneous matrix can be obtained and that the matrix is formed in one step. The disadvantages are that the high temperatures involved may result in reaction between molten matrix and reinforcement and that the high viscosity of molten ceramic often makes infiltration difficult. Due to differential shrinkage between matrix and reinforcement, the matrix is liable to crack on solidification. Refer figure 5.8

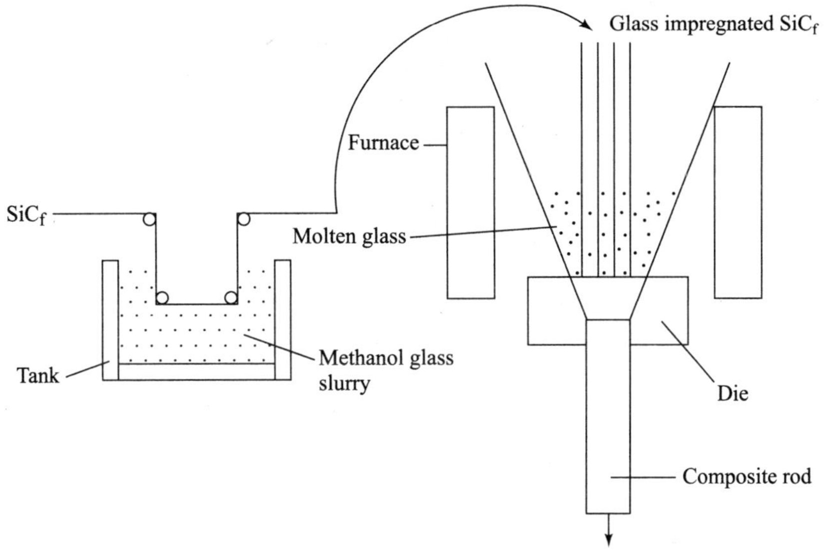

Figure 5.8 Pultrusion

5.2.2.2 Lanxide Process

This is a directed metal oxidation process. First, a preform is made. It will be a ceramic green body made by the usual method in case of a particulate composite. In case of a fibrous composite filament winding or a fabric lay-up is used to make the preform. A barrier is placed on the surface of the preform along all the sides to prevent growth of matrix material. The preform is then placed over the molten metal. The desired reaction product forms on the surface of the molten metal and grows outwards. The metal is supplied continuously at the reaction front by the wicking action through the channels in the oxidation product. Molten aluminium reacts with air to form aluminium oxide and with nitrogen to form aluminium nitride.

The chemistry of this process is difficult to control. Besides, this process cannot be used for large, complex parts. SiC_p/Al_2O_{3m} and ZrB_2 platelet/ZrC_m are produced by this technique. Refer figure 5.9

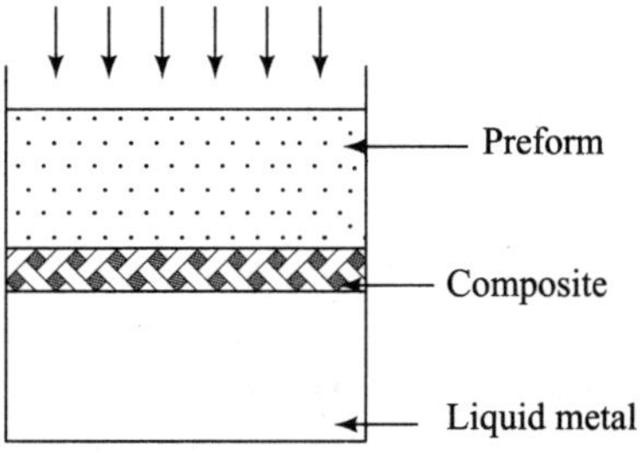

Figure 5.9 Directed metal oxidation (Lanxide) process

5.2.2.3 *Insitu chemical techniques*

(a) Chemical vapour impregnation: A preform made of yarn or woven fabrics or three-dimensional shapes is bombarded by the vapours of a matrix material such as SiC, Si_3N_4, or HfC produced by chemical vapour decomposition. The solid material produced by the gaseous reactants is deposited on the heated substrate, which is the preform.

This being a low temperature and low stress process, the problems associated with high temperature are eliminated. The other advantages are, (1) Composites formed by this process exhibit good mechanical properties at high temperatures. (2) Large complex shapes can be produced almost wholly (3) Suitable for oxide and non-oxide fibres and matrices. The disadvantages are that the process is very slow as it involves difussion of the reactant species in to the fibre substrate followed by outflow of the gaseous products. Besides, it is an expensive process. Refer figure 5.10. Si_3N_4 matrix can be deposited on a substrate of preform by using the following gasses such as silanes or siliceous halides with ammonia and reacting them between 800 – 1100°C.

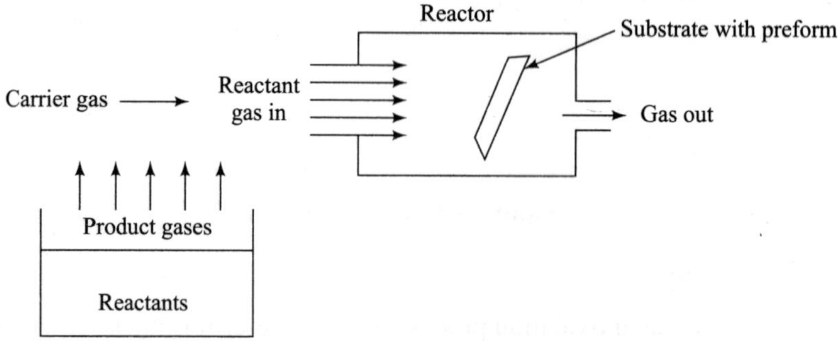

Figure 5.10 Chemical vapour infiltration

$3SiH_4(g) \longrightarrow 4NH_3(g)+Si_3N_4(s)+12H_2(g)$
$3SiCl_4(g) + 4NH_3(g) \longrightarrow Si_3N_4(s)+12HCl(g)$
$3SiHCl_3(g) + 4NH_3(g) \longrightarrow Si_3N_4(g)+9HCl(g) + 3H_2(g)$

For depositing a matrix of SiC methyltrichlorosilane is decomposed as follows at 1200-1400K.

$CH_3Cl_3Si(g) \longrightarrow SiC(g)+3HCl(g)$

For depositing alumina matrix, $AlCl_3$, H_2, CO_2 gasses are reacted at 1223-1273K at a pressure of 2-3 kPa.

$H_2(g)+CO_2(g) \longrightarrow H_2O(g) + CO(g)$
$2AlCl_3(g) + 3H_2O(g) \longrightarrow Al_2O_3(s)+6HCl(g)$
$2AlCl_3(g) + 3H_2(g)+3CO_2(g) \longrightarrow Al_2O_3(s)+3CO(g)+6HCl(g)$

5.2.2.4 Sol-gel techniques

A solution containing metal components (eg. a metal alkoxide, a metal acetate, or a metal halide) is reacted to form a sol. The sol is converted to a gel, which in turn is subjected to controlled heating to produce the desired end product - a glass, a glass-ceramic or a ceramic. The temperature required is quite low. High shrinkage and low yield are the main disadvantages of this process. Refer figure 5.11a&b. A sol is a colloidal suspension in which the individual particles are so small that no sedimentation occurs. The particles vary in size from 1nm to 100 nm. Sol has a free flowing form. It is obtained b chemical polymerisation of suitable compounds in the liquid state or by mixing a metal containing precursor (eg: a metal alkoxidie) a solvent (eg: an alcohol), a catalyst (eg: an acid) and water. Hydrolysis and polycondensation reactions make the sol more viscous until a gelled state is obtained. This gel is like a wet solid. It consists of a network of particles, alcohol & water. The wet gel is dried to remove any unwanted residue (water, organic compounds etc). A large amount of shrinkage occurs in this step. The porous gel is converted into a glass, ceramic or glass-ceramic by heating at low temperatures which are very much lower compared to direct melting of ceramic or powder metallurgy processes. This process had its beginning is 1846. Later in 1920 the chemistry became more understandable. To get silica glass, ceramic or glass-ceramic, starting material is $SiCl_4$ which reacts with ethyl alcohol to give $Si(OC_2H_5)_4$ called tetraethoxysilane. From metal alkoxide a metal oxide is formed. Polycarbosilane is the starting material for SiC matrix and amino silane for Si_3N_4 matrix production.

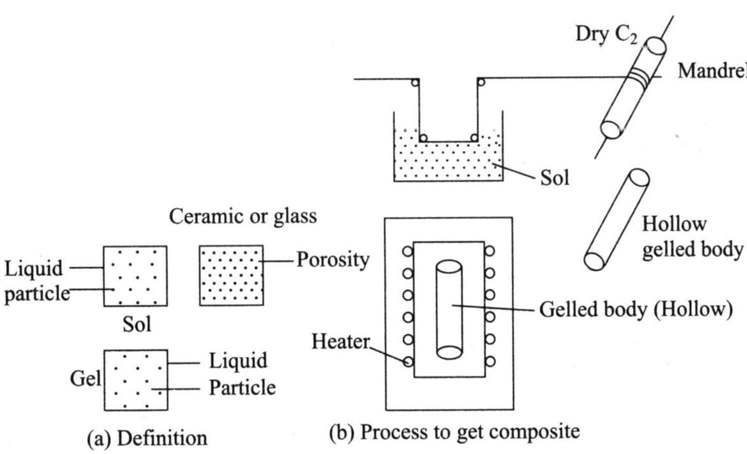

Figure 5.11 Sol - Gel process

5.2.2.5 Polymer pyrolysis

Covalent ceramics can be produced by pyrolysis of polymeric precursors at temperatures as low as 1400°C and with yields greater than those for chemical vapour impregnation processes. The advantages of this method are, (1) good matrix composition control (2) Relatively easier infiltration of fibre forms (3) relatively low densification temperatures. The disadvantages are, (1) low yields, (2) large shrinkage on sintering (3) requirement of multiple infiltration/densification cycles.

Porous SiC or Si_3N_4 fibrous preforms are evacuated in an autoclave and are infiltrated with molten precursors such as silaanes or polycarbosilanes at 500°C. Argon or nitrogen at a pressure of 2-40 MPa is passed. This results in the transformation of oligomer silane to polycarbosilane and simultaneous polymerisation at high pressures. Infiltrated samples are cooled and treated with solvents. The polymer is decomposed in an inert atmosphere at 800–1300°C and high pressure. These steps are repeated to obtain adequate density. The samples are finally annealed at 1030–1530°C to produce an optimum crystal structure. The composite, SiC_w /Al_2O_{3m} is produced by this technique. Refer figure 5.12

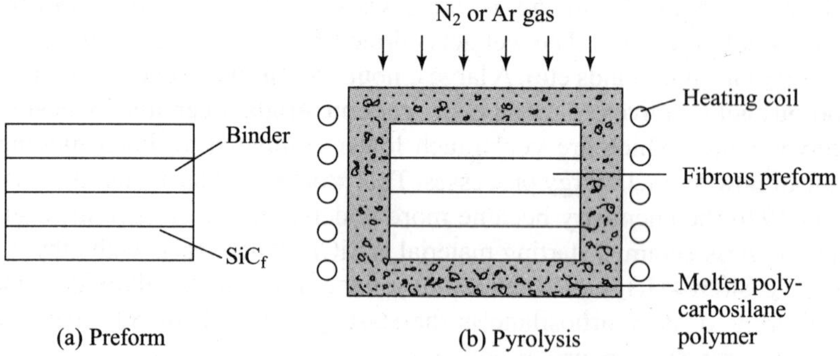

Figure 5.12 Polymer pyrolysis

5.2.2.6 *Self propagating high temperature synthesis (SHS)*

High combination temperature, simple low cost equipment and good control of process chemistry are the hallmarks of this process. Different shapes and forms can be obtained. No external source of energy is needed. Exothermic reactions are exploited to synthesize ceramic compounds, which cannot be produced by conventional techniques. An example is the composite, TiB_{2p}/$TiAl_m$ produced by the Exothermic Dispersion (XD) process. A mixture of the matrix and the reinforcement powder materials is heated. An exothermic reaction takes place. A metallic or intermetallic matrix with an intermetallic or ceramic reinforcement of particles or whiskers is obtained. Mix carbon, alumunium and titanium dioxide powder thoroughly and keep it in a cylindrical mould. Ignite the mixture at the bottom. The flame front passes from bottom to top causing the following reaction to take place.

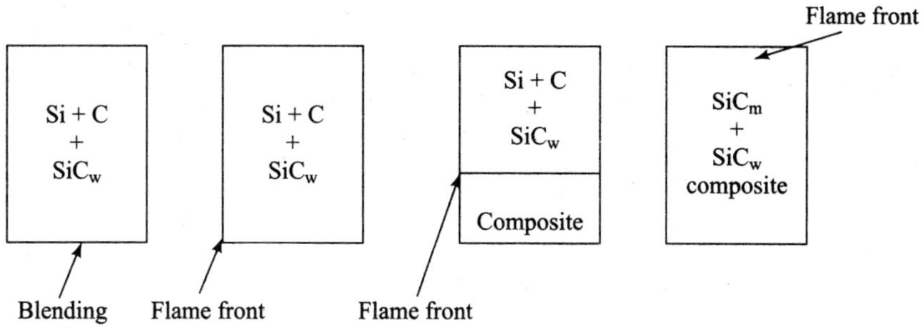

Figure 5.13 Self propagating High temperature Synthesis

$$4Al + 3TiO_2 + 3C \longrightarrow 2Al_2O_3 + 3TiC$$

We get a composite of alumina and titanium carbide. Similarly one can get SiC_w/ SiC_m composite also by effecting the following reaction $Si + C + SiC_w \rightarrow SiC_m + SiC_w$. Refer Fig. 5.13.

CHAPTER 6

CARBON-CARBON COMPOSITES (CCCs)

6.1 INTRODUCTION

Carbon - Carbon Composites (CCCs) consist of carbon or carbonaceous matrix reinforced with carbon fibres. There is a variety of carbon fibres. These may be produced from rayon, pitch or poly acrylo nitrile (PAN) and may be continuous, discontinuous chopped or woven. The matrix may be obtained by carbonisation of an organic material or by chemical vapour deposition or by a combination of these. Composites with a wide range of properties can be produced. Specific properties can be obtained by judicious, selection of matrix-fibre combination. Carbon carbon composites retain their strength even at temperatures at which metals and ceramics soften. They possess good thermal shock resistance due to their high thermal conductivity and low coefficient of thermal expansion. Besides they possess good abrasion resistance. They are biocompatible and have excellent wear resistance and chemical resistance. They have high heat of vaporisation and excellent fracture toughness.

CCCs constitute a unique, family of composite materials that finds widespread application in high temperature aerospace equipments and advanced engineering products such as race cars, superfast trains and submarines. However, as on date, the production processes are inefficient and expensive. Secondly, the oxidation resistance of CCCs at elevated temperatures is poor. Usually, protection has to be provided in the form of a coating of SiC or Si_3N_4. Since this, primary coat is liable to crack in service, a secondary coat of glass containing B or Si is called for. At high temperatures, B or Si will seal the cracks that develop. CCCs are now mainly used in military applications where high cost of production is compensated for, by specific performance. Thus, as of now, CCCs are super speciality materials and will continue to be so for some more time. CCCs are being used in the manufacture of brakes and clutches, rocket motors, heat shields for re-entry space vehicles, engine components of aircraft, hot forming dies, furnace heating elements, high temperature fastners, tennis racquets, etc.

6.2 PRODUCTION

There are 3 different routes as given below for producing CCCs

(1) Chemical vapour deposition (CVD) technique such as the isothermal method, thermal gradient method, pressure gradient method and pulse CVD.
(2) Pyrolysis using thermosets
(3) Pyrolysis using thermoplastics

6.2.1 Chemical Vapour Deposition

The major techniques, which involve chemical vapour deposition, are discussed hereunder.

6.2.1.1 Isothermal method : A susbtrate is placed in an even temperature furnace. The reactant gases are passed over it so that diffusion takes place. To prevent the sealing of pores at the mouth, the surface reaction rate is kept lower than the diffusion rate. This is very significant since it implies slow weight gain and long processing time. Full densification is impossible since the amount of porosity decreases with time. A compromise between densification and deposition rate has to be struck. The isothermal method is simple and allows for using large furnaces to impregnate several objects at a time. The main disadvantage is that the processing time is large in spite of the low deposition pressure and diluent gases employed. Refer figure 6.1

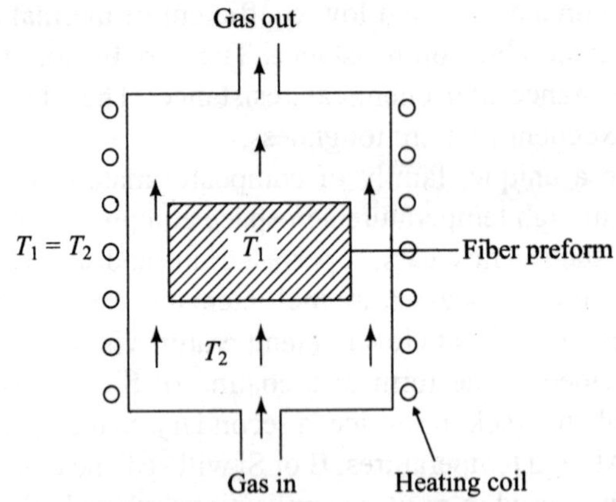

Figure 6.1 Isothermal method

6.2.1.2 Thermal gradient method : It is a diffusion controlled process. A temperature gradient is maintained across the fibre preform so that over crusting is avoided. The fibre preform is wound on a mandrel made of graphite which acts as an inductive heating susceptor. Temperature, thermal conductivity of perform and gas flow rate are the variables which influence deposition at the interface of fibre and mandrel. Deposit grows outwards from the surface of the fibre preform. The fibre preform should have low thermal conductivity and the gas flow rate must be high.

The advantages of the method are: (1) high deposition rates as compared to conventional isothermal method. (2) over crusting is reduced (3) atmospheric pressure can be used (4) involves no machining of samples unlike the conventional method.

The disadvantages are (1) limited to one item at a time (2) scaling up is difficult. Refer figure 6.2

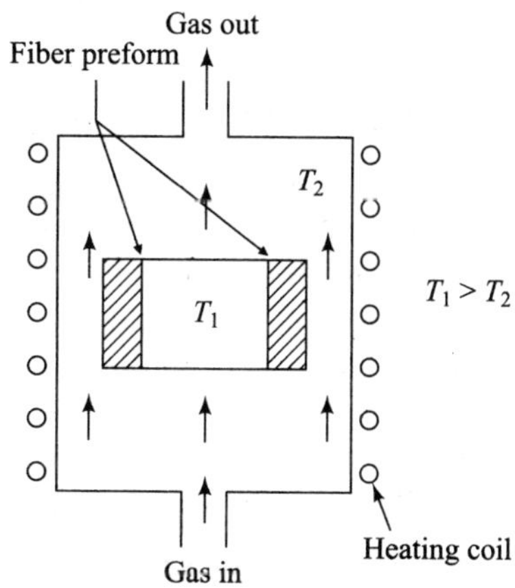

Figure 6.2 Thermal gradient method

6.2.1.3 Pressure gradient method : This method is as follows: A fibre preform is placed in a gas tight unit and heated. The resistance to gas flow of the fibre preform causes a pressure gradient to be set up. Unlike the previous two processes, here the deposition rate increases as the pressure gradient rises when the pores close up due to prior deposition. Rate of deposition is directly proportional to pressure drop across the preform. Refer figure 6.3

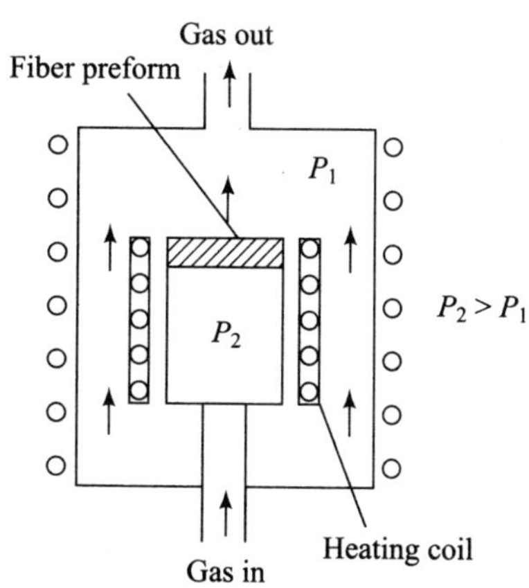

Figure 6.3 Pressure gradient method

6.2.1.4 Pulse CVD : Here a hot reaction chamber is alternatively subjected to vacuum and atmospheric pressures so that the reactant gases are forced deep into the pore structure of the preform. In every cycle the unwanted products are pumped out by the fresh reactant gases and the pores are ready to be infiltrated in the next cycle. This method completely overcomes the limitations imposed by diffusion. However, it is not commonly used due to the complication of cycling the reaction chamber rapidly between atmospheric pressure and near vacuum of a few torrs in every cycle.

6.2.2 Pyrolysis Using Thermosets

Here a laminated structure is pyrolysed in an inert atmosphere to around 1000°C. The resulting product is heated to 2500°C to bring about graphitisation of the carbon matrix. This is an extremely inefficient process. Nevertheless, it is inexpensive and quick and imposes no size limitation. Techniques such as hand lay up, pultrusion, prepregging, filament winding or compression moulding may be used to obtain the laminated structure.

6.2.3 Pyrolysis Using Thermoplastics

Hot isostatic pressure impregnation carbonisation often referred to as the HIPIC process is used to form carbon-carbon composites. It involves the following steps: (1) A dry fibre preform or porous laminate is vacuum impregnated with molten pitch. (2) It is placed in a metal can or container and surrounded by excess pitch. (3) The can is evacuated and sealed (4) It is then placed in a Hot Isostatic Pressure (HIP) unit. (5) Temperature is raised above the melting point of pitch. (6) Pressure is increased to 100 MPa. Pitch melts, expands and is forced into the pores in the sample. The sealed container acts like a rubber bag and helps in carrying the pitch to the interior portions of the sample through the pores. Following this, carbonisation takes place. The pressure applied increases the yield of carbon from pitch. (7) After processing, the preform is removed from the container. (8) It is cleaned. (9) It is graphitised by heating to 2300°C. Steps 1 to 9 are repeated till the required density is obtained.

CHAPTER 7

INTERMETALLIC MATRIX COMPOSITES (IMCs)

7.1 INTRODUCTION

Intermetallic phases are ordered and disordered compound phases of two or more metals. Pearson's Handbook of Crystallographic Data of Intermetallic Phases includes thousands of compounds and runs into four volumes of several hundred pages each. Aluminides of titanium, nickel and iron are examples of such compounds containing intermetallic phases. Intermetallic composites consist of a matrix of an intermetallic compound reinforced with fibres, whiskers or particles.

7.2 INTERMETALLIC COMPOSITES

The properties of intermetallic composites may be attributed to the individual properties of the matrix material and the reinforcement as well as the effect of combining them.

7.2.1 Matrix Materials

Ordered intermetallics have a number of properties that make their composites suitable for high temperature applications. Some such properties are, high melting points, adequate resistance to environmental degradation, structural and chemical stability at high temperatures and retention of specific mechanical properties at elevated temperatures. IMCs are of special significance in aerospace component manufacturing due to the substantial reduction in weight possible by their use. The more promising and widely investigated among the intermetallic matrices are the aluminides of titanium, nickel and iron. Their properties and crystal structures are given in Table 7.1 and their unit cells are shown in Fig 7.1. The aluminides however, are restricted to use well below 1600°C. They are usually used between 820-1260°C For turbines that can meet the demands of the future and for compressors to be used in aerospace engines, silicide matrices that can withstand up to 1600°C may be required. $MoSi_2$ and Ti_3Si_3 are the more significant among these.

Table 7.1 Properties of Aluminides

Intermetallic Compound	Crystal Structures	M.P. °C	Critical temperature T_c	Density, kg/m^3 X10^3	Young's Modulus, GPa
FeAl	(B_2) Ordered BCC	1250	1250	5.6	263
Fe_3Al	(DO_3) ordered BCC	1540	540	6.7	141
	(B_2) Ordered BCC	1540	760	6.7	141
NiAl	(B_2) Ordered BCC	1640	1640	5.9	294
Ni_3Al	(Ll_2) Ordered FCC	1390	1390	7.5	178
TiAl	(Ll_o) Ordered FCT (c/a) = 1.02	1460	1460	3.9	94
Ti_3Al	(DO_{19}) Orderd CPH c = 0.420, a = 0.577	1600	1180	4.2	210

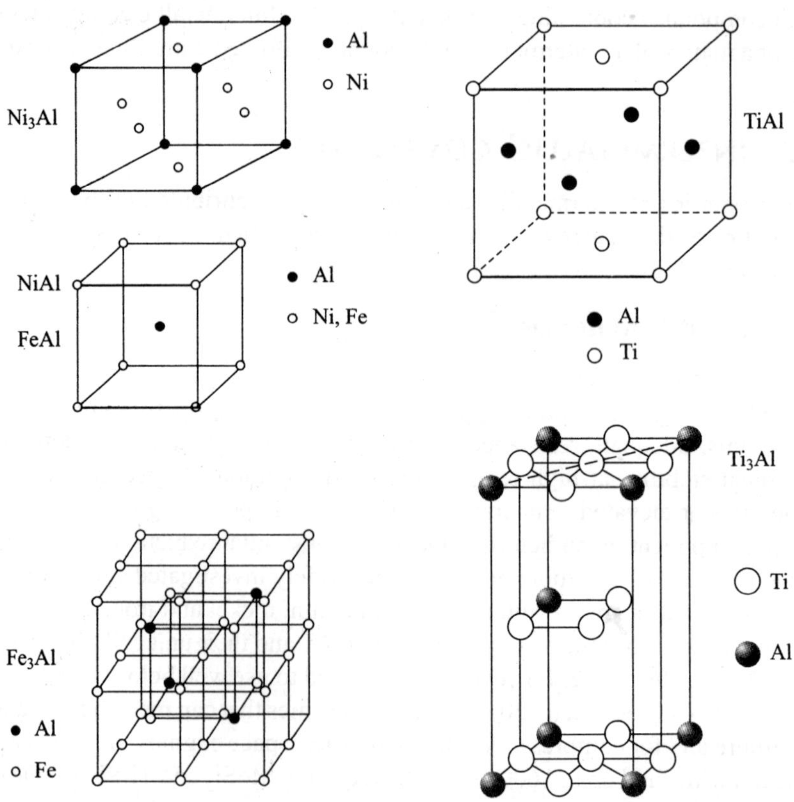

Figure 7.1 Crystal structure of aluminides of nickel, iron and titanium as represented by their unit cells

They possess high strength, creep resistance, oxidation resistance and microstructural stability. As such, silicates tend to be brittle below 800°C and plastic above 800°C. However, silicate-based composites with particle, whisker or fibre reinforcement are tough at low temperatures and creep resistant at high temperatures. The possibilities of beryllide and chromide matrices too have been investigated. Examples are niobium beryllide, zirconium beryllide, titanium chromide and niobium chromide.

7.2.2 Reinforcements

These could be in the form of particles, whiskers or fibres. SiC, Al_2O_3, B, C, metals etc. are some of the reinforcements used. Intermetallic matrices should be chemically, physically and mechanically compatible with reinforcements. Intermetallics are reinforced to increase their toughness. When used without reinforcement they are brittle.

7.3 PROCESSING AND PRODUCTION TECHNIQUES FOR IMCs

Various techniques have been developed for the processing and prodction of IMCs.

7.3.1 Power Metallurgy

Blends of fine alloy powder are mixed with water to from a slurry and this slurry is frozen into sheets. These are then interlayered with reinforcing fibres. Vacuum degassing is done at 400°C. Water is removed. Heat and pressure are applied to consolidate the laminated structure. Examples of composites prepared using this technique are Ti_3Al_m / SiC_f and Al_2 Ta_m / Al_2O_{3f}

7.3.2 Exothermic Dispersion (XD)

It is based on high temperature synthesis. Elemental borax, titanium and aluminium powder are blended compacted and then ignited. As the flame front passes through, the reactants, titanium aluminide and titanium diboride are formed. The resultant material contains 20 to 75% TiB_2. This can be used as a master alloy and added to TiAl in required quantities.

7.3.3 Liquid Melt Infiltration

Disk or rod shaped composites can be produced by this technique. For making disks, an intermetallic alloy is melted and made to infiltrate a mould containing fibres which are also hot. Induction heating is applied separately on the alloy and the fibres in vacuum to a temperature just above the melting point of the alloy. Rods are produced by inserting an Al_2O_3 tube containing the fibres into a crucible holding the alloy to be melted. Heating is done in vacuum to a temperature just above the melting point of the alloy. The tube is fully immersed into the molten alloy and the molten alloy is pushed into the tube to infiltrate the fibres.

7.3.4 Powder Cloth Process

The intermetallic alloy powder is mixed with Teflon powder, the binder and high grade kerosene (stoddard solution). The mixture, after the kerosene evaporates, becomes dough, which is rolled into thin sheets and trimmed to size after evaporating the remaining kerosene. This sheet is alternated with fibre mats to obtain the required number of layers. This is followed by vacuum hot pressing. All the binder has to be eliminated in this step for achieving the desired properties.

7.4 PROMISING IMCs SYSTEMS

Following is a list of systems that have been thoroughly investigated and found to be promising for future applications. Subscripts used have the following notation: m-matrix p-particle f-fibre w-whisker.

1. Ti_3Al_m/SiC_f
   [Eg., $(Ti\text{-}25\ Al\text{-}10Nb)_m$ / SiC_f, $(Ti\text{-}24\ Al\text{-}11Nb)_m$ / SiC_f]
2. $TiAl_m/Al_2O_{3f}$
3. $TiAl_m$/TiB2p
4. Ni_3Al_m/Al_2O_{3f}
5. $NiAl_m/Al_2O_{3f}$
6. $NiAl_m/TiB_{2p}/Al_2O_{3f}$
7. $NiAl_m/B_4C_p$/304 stainless steel tubes (reinforcement)
8. $FeAl_m/Al_2O_{3f}$
9. $FeAl_m/(ZrO_2\text{-}Al_2O_3)_f$
10. Fe_3Al_m/Al_2O_{3p}
11. $(NbAl_3/Nb_2Al)_m+(ZrO_2\text{-}Al_2O_3)_f$

7.5 PROPERTIES OF IMCs

Aluminide IMCs generally possess the following properties:

1. Superior tensile properties as compared to super alloys
2. Good oxidation resistance
3. High melting points
4. Low chemical compatibility between available fibres and intermetallic matrix materials
5. Enormous mismatch in coefficient of thermal expansion (CTE) of fibres and matrices
6. Follows Rule of Mixtures (ROM)

Deformation behaviour remains to be investigated thoroughly. Not much is known about the behaviour of the matrix and the fibre when in combination. Individually the matrix may deform plastically within the composite.

7.6 TOUGHNESS OF IMCs

Toughness of IMCs can be improved by

1. Marco alloying and crystal structure modification – tracers of Nb may be used in Ti_3Al
2. Micro alloying - Be is added to Ni_3Al and B to Ni_3Al
3. Controlling grain size or shape – as in NiAl
4. Reinforcement by ductile fibres or particle - Nb reinforcement in TiAl
5. Controlling substructure
6. Applying hydrostatic pressure – as in Ni_3Al
7. Inducing martensitic transformation - Fe is added to NiAl for this.
8. Use of alloy Tubes to reinforce matrix – such as 304 stainless steel tubes in NiAl.

7.7 APPLICATIONS OF IMCs

In the near future, IMCs are bound to find favour with aerospace applications. They are suitable for manufacturing.

1. Compressors, HP Turbines and LP Turbines of subsonic aircraft
2. Compressors for commuter rotorcraft
3. Compressors and nozzles for high speed aircraft.

CHAPTER 8

MECHANICS OF COMPOSITE MATERIALS

8.1 CONTINUOUS FIBRES

Composites follow the Rule Of Mixtures (ROM) which can be derived for the cases of isostress and isostrain conditions. See Fig.8.1 (a) and (b) given below.

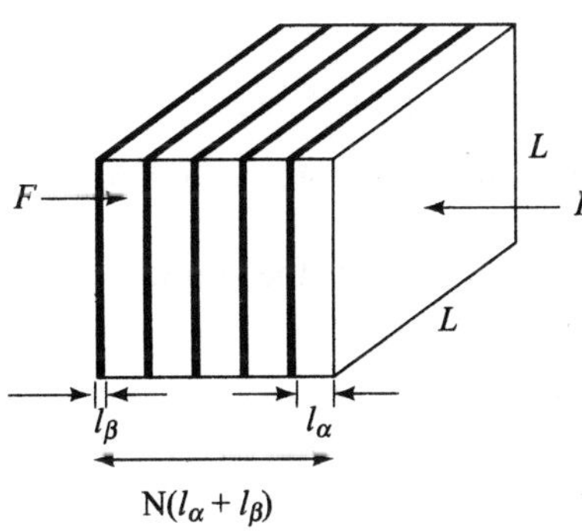

Figure 8.1 (a) Isostress condition

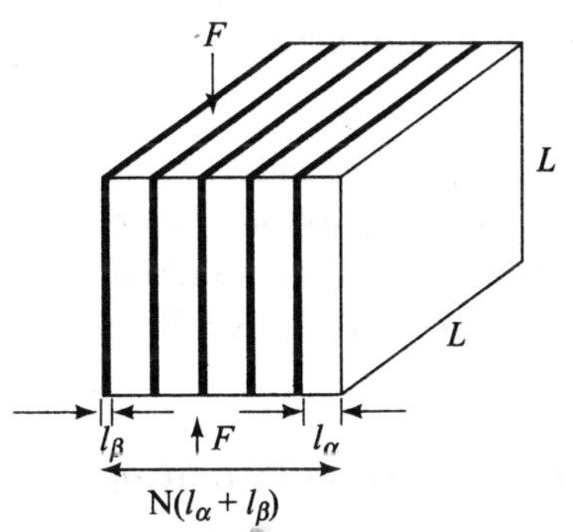

Figure 8.1 (b) Isostrain condition

Consider a two phase aggregate of cross section *LxL* and length $N(1_\alpha + 1_\beta)$ where 1_α is the length of one α layer and 1_β is the length of one β layer and there are 'N' α layers arranged alternately with 'N' β layers.

8.1.1 Isostress Condition

F is the force applied normal to the *LxL* cross-section. This force causes both, the α layer and the β layer to elongate. If $\Delta 1_\alpha$ and $\Delta 1_\beta$ denote the elongation for each α and β layer, ε_α and ε_β the strains in these layers, we have,

$$\Delta 1_\alpha = \varepsilon_\alpha 1_\alpha \qquad (1)$$

$$\Delta 1_\beta = \varepsilon_\beta 1_\beta \qquad (2)$$

For N layers the elongation is given by

$$N (\Delta 1_\alpha + \Delta 1_\beta)$$

Therefore elongation of the composite

$$\Delta 1_c = N (\Delta 1_\alpha + \Delta 1_\beta) \qquad (3)$$

If F is the force applied as shown the stress is given by F/L^2 and it is the same for an α layer, β layer and the overall composite in case of iso stress condition. Thus, if σ_C, σ_α and σ_β are the respectively the stress for the composites an α layer and a β layer, we have,

$$\text{i.e., } \sigma_c = \sigma_\alpha = \sigma_\beta = (F/L^2) \tag{4}$$

Strain of composite $\varepsilon_C = [\Delta l_c / N(l_\alpha + l_\beta)]$

$$[N(\Delta l_\alpha + \Delta l_\beta)] / [N(l_\alpha + l_\beta)] = [\Delta l_\alpha / (l_\alpha + l_\beta)] = [\Delta l_\beta / (l_\alpha + l_\beta)] \tag{5}$$

Volume fraction of α

$$V_\alpha = l_\alpha / (l_\alpha + l_\beta) \tag{6a}$$

Volume fractions of β

$$V_\beta = (l_\beta)/(l_\alpha + l_\beta) \tag{6b}$$

Substituting these in equation (5)

$$\varepsilon_c = (\Delta l_\alpha V_\alpha) / (l_\alpha) + (\Delta l_\beta V_\beta) / (l_\beta)$$

$$= \varepsilon_\alpha V_\alpha + \varepsilon_\beta V_\beta \tag{7}$$

For the case of elastic deformation

$$\varepsilon_\alpha = \sigma / E_\alpha, \quad \varepsilon_\beta = \sigma / E_\beta, \quad \varepsilon_c = \sigma / E_c$$

where, E_α, E_β and E_c are respectively, the values of young's modulus for the α,β and the composites layers.

Substituting these values of ε_α, ε_β and ε_c in (7) one gets the young's modulus of the composite E_c as follows

$$E_c = [(E_\alpha E_\beta)/(V_\alpha E_\beta + V_\beta E_\alpha)] \tag{8}$$

8.1.2 Isostrain Condition

For the isostrain condition, $\varepsilon_c = \varepsilon_\alpha = \varepsilon_\beta$ and F is applied on the area $L \times N (l_\alpha + l_\beta)$ (9)

$$F_\alpha = \sigma_\alpha N l_\alpha L \tag{10}$$

$$F_\beta = \sigma_\beta N l_\beta L \tag{11}$$

$$F = NL(\sigma_\alpha l_\alpha + \sigma_\beta l_\beta) \tag{12}$$

$$F/NL(l_\alpha + l_\beta) = \sigma_c = (\sigma_\alpha l_\alpha)/(l_\alpha + l_\beta) = (\sigma_\beta l_\beta)/(l_\alpha + l_\beta) \tag{13}$$

Therefore

$$\sigma_c = \sigma_\alpha V_\alpha + \sigma_\beta V_\beta \tag{14}$$

Normally composites behave in isostrain fashion and equation (14) is widely used to predict their mechanical behaviour. In the case of elastic deformation, the modulus of the composite is calculated as follows.

$$E_c = \sigma_c / \varepsilon_c \; E_\alpha = \sigma_\alpha / E_\alpha \text{ and } E_\beta = \sigma_\beta / E_\beta \tag{15}$$

Substituting these in equation (14) we obtain the value of the modulus of the composite as follows.

$$E_c = V_\alpha E_\alpha + V_\beta E_\beta \tag{16}$$

8.1.3 Stress Vs Strain

Stress-strain behaviour of continuous fibre composite comprises of 3 stages, assuming isostrain behaviour of the composite. If V_f and V_m denote the volume fraction of fibre and matrix respectively, the equations for the three stages would be,

Stage 1 : Fibre is elastic and matrix is elastic

$$\sigma_c = E_c\, \varepsilon_c = \varepsilon_c\, [V_f E_f + V_m\, E_m] \tag{17}$$

Stage 2 : Fibre is elastic and matrix is plastic

$$\sigma_c = V_f E_f \varepsilon_c + V_m\, \sigma_m(\varepsilon_c) \tag{18}$$

Stage 3 : Fibre is plastic and matrix is plastic

$$\sigma_c\,(\varepsilon_c) = V_f\, \sigma_f(\varepsilon_c) + V_m\, \sigma_m\,(\varepsilon_c) \tag{19}$$

$$\text{Since } \varepsilon_c = \varepsilon_f = \varepsilon_m \tag{20}$$

8.1.4 Critical Volume Fraction of Fibre

If T.S denotes the tensile strength and f, m and c have the usual implications, we may write,

$$\mathrm{T.S_c} = \mathrm{V_f\,(T.S.)_f} + V_m\, \sigma_m\,(\varepsilon_f) \tag{21}$$

Where, $(\mathrm{T.S})_c$ – composite tensile strength
$\mathrm{T.S_f}$ - Fibre tensile strength
$\sigma_m\,(\varepsilon_f)$ - Matrix flow stress at fibre strain ε_f
If V_c represents the critical volume fraction of fiber,

$$(\mathrm{T.S})_c = V_m\,(\mathrm{T.S})_m \text{ when } V_f \leq V_c \tag{22}$$

$$(\mathrm{T.S})_c = V_f(\mathrm{T.S})_f + V_m\, \sigma_m\,(\varepsilon_f) \text{ when } V_f > V_c \tag{23}$$

From equations (22) and (23), for critical volume fraction of fibre, the following equation holds

$$V_m\,(\mathrm{T.S.})_m = V_f\,(\mathrm{T.S,})_f + V_m\, \sigma_m\,(\varepsilon_f) \tag{24}$$

$$V_m\,[(\mathrm{T.S})_m - \sigma_m\,(\varepsilon_f)] = V_f\,(\mathrm{T.S})_f$$

$$(1\text{-}\mathrm{V_f})\,[(\mathrm{T.S})_m - \sigma_m\,(\varepsilon_f)] = V_f\,(\mathrm{T.S})_f$$

$$(\mathrm{T.S})_m - \sigma_m\,(\varepsilon_f) = V_f\,[(\mathrm{T.S})_f + (\mathrm{T.S_m} - \sigma_m\,(\varepsilon_f)]$$

Thus the critical volume fraction of fibers V_c is given by

$$V_f = V_{cri} = [(\mathrm{T.S})_m - \sigma_m\,(\varepsilon_f)] / \,[(\mathrm{T.S})_f + (\mathrm{T.S})_m - \sigma_m\, \varepsilon_f] \tag{25}$$

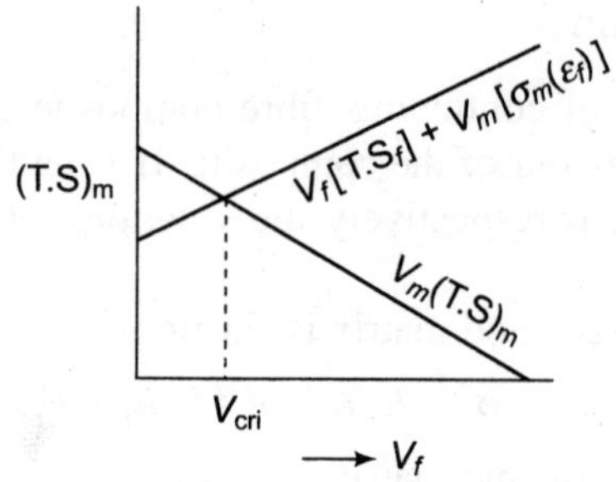

Figure 8.2 (a) Critical volume fraction of fibre

8.1.5 Minimum Volume Fraction of Fibre

For minimum volume fraction of fibre, we may assume that $(T.S.)_c$ is equal to $(T.S)_m$ so that equation (23) may be written as,

$$(T.S)_m = V_f (T.S)_f + V_m\, \sigma_m\, (\varepsilon_f) \quad (26)$$

$$= V_f (T.S)_f + (1 - V_f)\, \sigma_m\, (\varepsilon_f)$$

$$= V_f (T.S)_f + \sigma_m\, (\varepsilon_f) - V_f\, \sigma_m\, (\varepsilon_f)$$

$$V_f [(T.S)_f - \sigma_m\, (\varepsilon_f)\,] = [(T.S)_m - \sigma_m\, (\varepsilon_f)]$$

Therefore

$$V_{min} = [(T.S)_m - \sigma_m\, (\varepsilon_f)] \,/\, [T.S)_f - \sigma_m\, (\varepsilon_f)] = V_f \quad (27)$$

Refer Fig 8.2 (b)

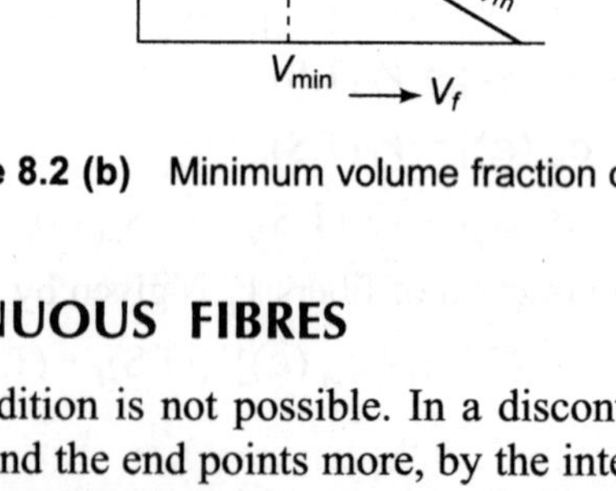

Figure 8.2 (b) Minimum volume fraction of fibre

8.2 DISCONTINUOUS FIBRES

Here equal strain condition is not possible. In a discontinuous fibre, the mid point is strained less and the end points more, by the interfacial shear stress. A force balance can be considered to exist between the interfacial shear force and the tensile force in the fibre as represented by the following equation.

$$\tau_m\, (\pi d_f)\, dx = [(\pi d_f^2)/4]\, d\sigma_f \quad (28)$$

where, τ_m-shear stress in the matrix, d_f– diameter of the fibre x-length of the fibre and σ_f– stress in the fiber. See Figs 8.3 (a) (b) (c) and (d).

Therefore $$(d\sigma_f/dx) = (4\tau_m/d_f)$$

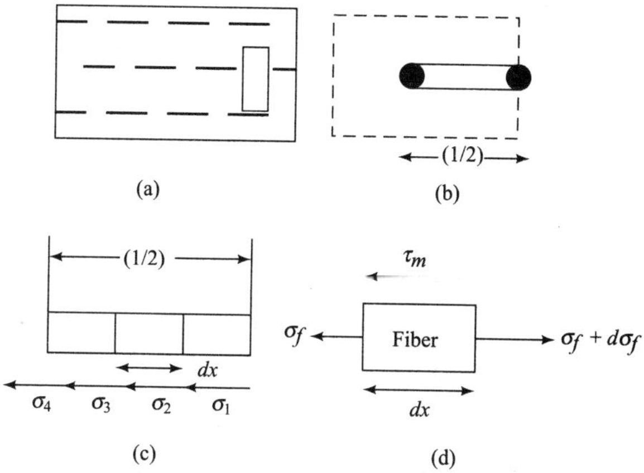

Figure 8.3 Stress distribution in discontinuous fiber

σ_f is zero at fibre ends and maximum at fibre mid point while τ_m is zero at fibre mid point and maximum at fibre ends when both, the matrix and the fibre are undergoing elastic deformation as shown in Fig. 8.4 (a). If the matrix is ideally plastic, the variation is as shown in figure 8.4 (b). There is a length l_c called the critical length of fibre, below which the fibre does not take the loading corresponding to composite strain and above which it takes the loading corresponding to composite strain over a stretch of fibre length in the middle. Refer Fig. 8.4 (c) At exactly the critical length of fibre the composite strain will exist only at the mid point of the fibre. The composite strain corresponds to the equal strain condition of fibre, matrix and composite [i.e., $\varepsilon_c = \varepsilon_f = \varepsilon_m$]. When $x = l_c/2$, stress is $\sigma_{f\varepsilon c}$.

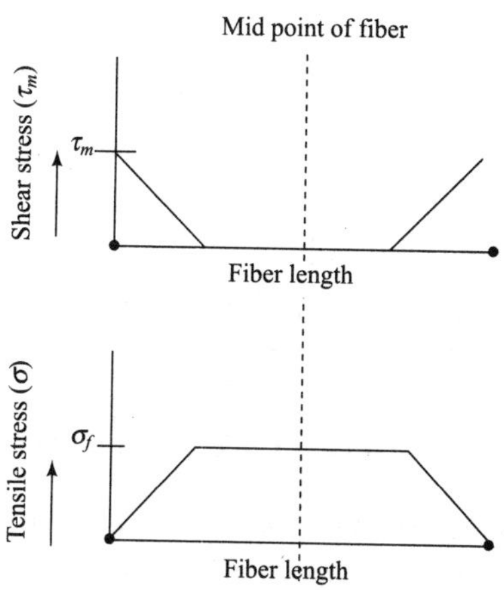

Figure 8.4 (a) Stress in discontinuous fiber when fiber is elastic and matrix is elastic

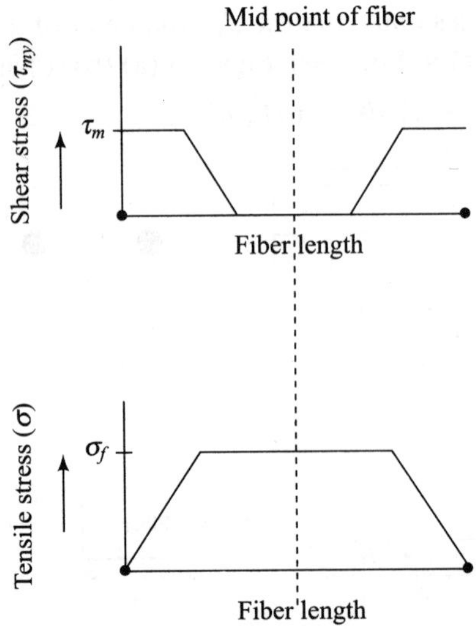

Figure 8.4 (b) Stress in discontinuous fiber when fiber is elastic and matrix is plastic

Therefore
$$\sigma_{f\varepsilon c} = 2\,\tau_{my}\,(l_c/d_f)$$
$$(l_c/d_f) = (\sigma_{fec}/2\,\tau_{my})$$

(l_c/d_f) is called the critical aspect ratio. Discontinuous fibre cannot take up the strain that a continuous fibre can. VFR for equal strain condition is not valid for such a situation. It gets modified as follows :

Situation-I : Length of fibre is greater than the critical length Refer 8.5 (a)

$$l > l_c$$

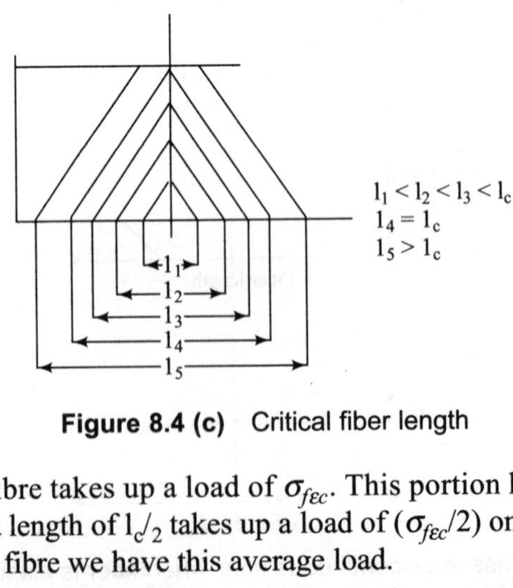

Figure 8.4 (c) Critical fiber length

A portion of fibre takes up a load of $\sigma_{f\varepsilon c}$. This portion has length equal to $(l-l_c)$. However, a length of $l_c/_2$ takes up a load of $(\sigma_{f\varepsilon c}/2)$ on the average and at the two ends of a fibre we have this average load.

$$\text{Total load} = \sigma_{f\varepsilon c}\,[\,(1\text{-}1_c)/1]\,[(\pi.d_f^2)/4] + (1_c/21)\,\sigma_{f\varepsilon c}\,(\pi d_f^2/4)$$

Therefore average stress becomes

$$\sigma_f = \sigma_{f\varepsilon c}\,[1 - (1_c/1)] + [(1_c/21)\,\sigma_{f\varepsilon c}]$$
$$= \sigma_{f\varepsilon c}\,[1 - (1_c/21)]$$

VFR becomes

$$\sigma_{c\varepsilon c} = V_f\,\sigma_{f\varepsilon c}\,[1 - (1_c/21)] + V_m\,/\sigma_{m\varepsilon c} \qquad (29)$$

where $\sigma_{m\varepsilon c}$ represents stress in matrix when ε is the strain in composite

Situation - II : Fibre length is lesser than the critical length. Refer Fig 8.5 (b)

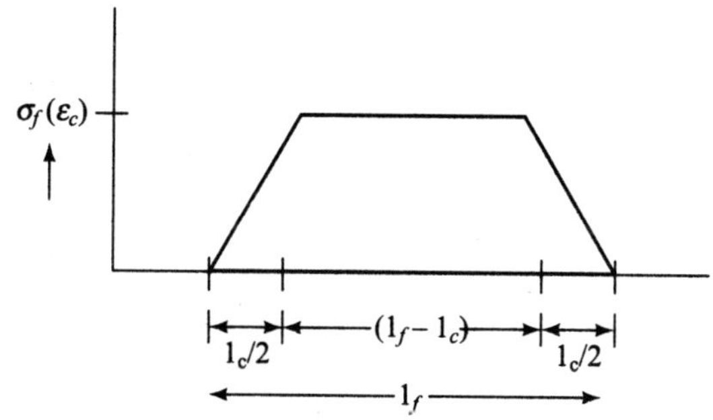

Figure 8.5 (a) Stress in length of fiber greater than critical length

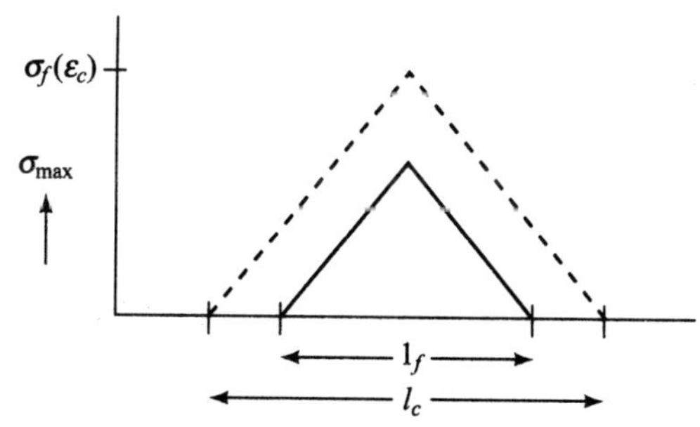

Figure 8.5 (b) Stress in length of fiber less than critical length

$$1 < 1_c$$

$$\text{Average stress} = (\sigma_{max}/2) = 2\tau_{my}\,1_c/d_f$$

Where σ_{max} is maximum stress and τ_{my} is the shear yield of matrix.

$$\text{But } \tau_{my} = \sigma_{f\varepsilon}\,(d_f/21_c)$$

Substituting for τ_{my} we get

$$\sigma_f = (1/2)\,\sigma_{f\varepsilon c}\,(1/1_c)$$

VFR becomes

$$\sigma_{c\varepsilon c} = V_f\,\sigma_{f\varepsilon c}\,(1/21_c) + V_m\,\sigma_{m\varepsilon c} \qquad (30)$$

Where $\sigma_{c\varepsilon c}$ denotes the stress in the composite when the strain in the composite is ε. The capability of a discontinuous fibre reinforced composite is lesser than that of continuous fibre reinforced composite. When $l = l_c$

$\sigma_{c\varepsilon c} = V_f(\sigma_{f\varepsilon c}/2) + V_m\ \sigma_{m\varepsilon c}$. This can be obtained by substituting in either equation 29 or 30.

8.3 NATURE OF STRESS Vs STRAIN CURVE FOR COMPOSITES

8.3.1 CMCs

There are two possibilities as given below:

Case 1 Fibre is ceramic and matrix is also ceramic. In the ideal case both will be elastic in nature. Then the stress strain curve will be linear. Fibre may break before matrix or matrix may break before fibre. At that point there will be slope change in the linear curve. Between first fiber breakage and composite failure there will be several serrations.

Case 2. Fibre is metallic and matrix is ceramic. In this case there will be two stages in the stress vs strain curve. In the first stage, the matrix and the fibre will be elastic. Therefore it will be linear. In the second stage the matrix will be elastic and the fibre will be plastic. Therefore this portion will be nonlinear.

8.3.2 MMCs

There are two possibilities here also.

Case 1. Matrix is metallic and fibre is ceramic. There will be two stage stress vs strain curve. In stage 1, both fibre and matrix are elastic and therefore linear. In stage two, the fibre is elastic but the metal is plastic and therefore the curve is non-linear.

Case 2. Matrix is metallic and fibre is also metallic. In this case there will be three stages. In stage 1, both fibre and matrix arte elastic and the curve is linear. In stage two fibre is elastic and matrix is plastic and the curve is nonlinear. In stage three both fibre and matrix are plastic the curve is non-linear.

8.3.3 PMCs

Here also there are two possibilities since the matrix may be thermoplastic or thermosetting plastic. For thermoplastic polymer matrix composites the stress strain curve has two stages. In the first stage, both fibre and matrix are elastic. In the second stage, the matrix is plastic and fibre is still elastic. This is due to thermoplastics showing plastic deformation.

For thermosetting polymer matrix composites there is only one stage in the stress strain curve. Both fibre and matrix are elastic. This is because thermosets do not show appreciable plastic deformation. Therefore the curve will be linear.

8.3.4 Stress – Strain Curves

Stress vs strain curve for a fiber is shown in Fig. 8.6 (a). Mostly fibers will be elastic. Fig. 8.6 (b) shows stress vs strain curve for ferrous and non ferrous metals. Fig. 8.6 (c) and (d) show such curves for ceramic matrix and polymer matrix. (Thermoset, Thermoplastic and Rubber). Figs 8.7 (a) (b) and (c) show stress vs strain curve for MMCs, CMCs and PMCs. Fiber failure is not synonymous with composite failure. Even after a fiber fails matrix carries the load. See Figs 8.8 and 8.9. For a metal matrix and metal fiber system the three stages in stress vs strain curve as for fiber, and matrix are compared in Fig. 8.10.

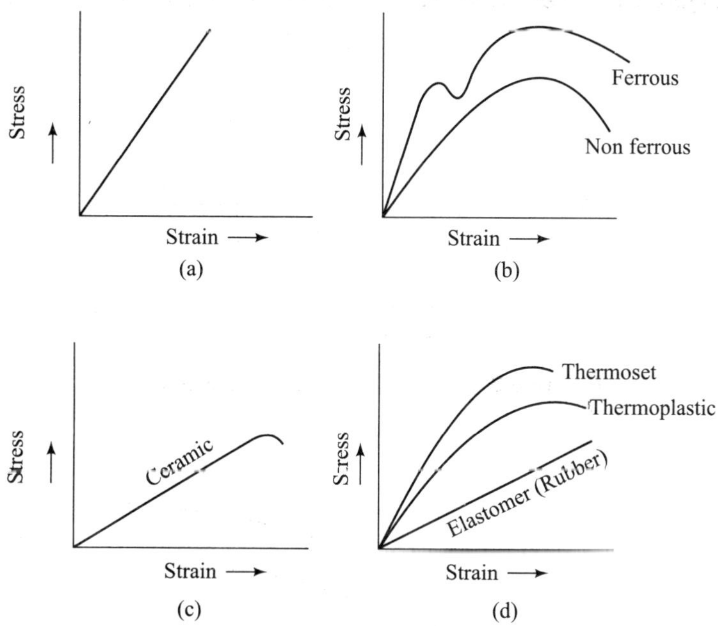

Figure 8.6 Stress against strain for reinforcement and matrices (schematic)

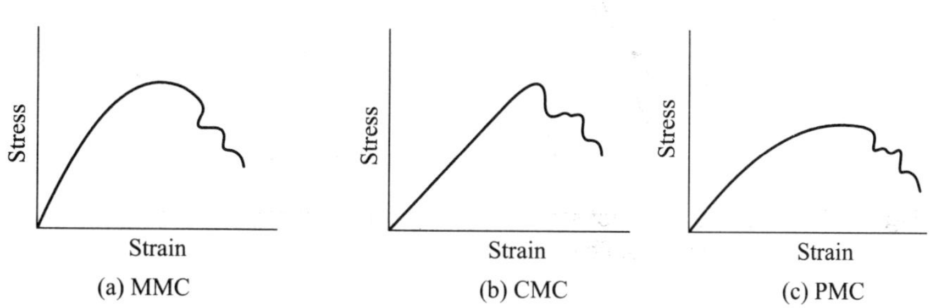

Figure 8.7 Variation of stress against strain for composites (schematic)

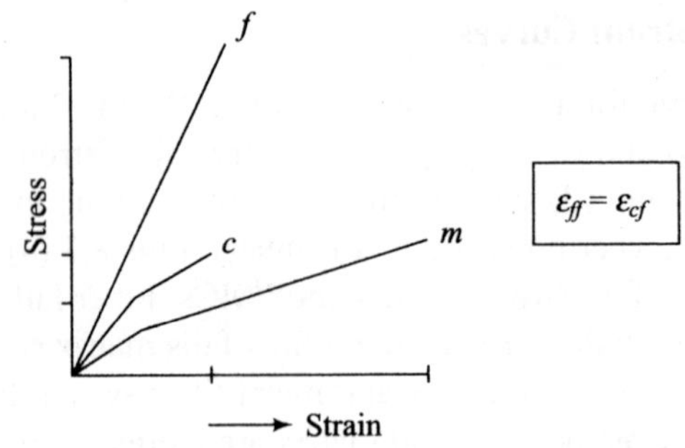

Figure 8.8 Fiber failure and composite failure occur at same strain

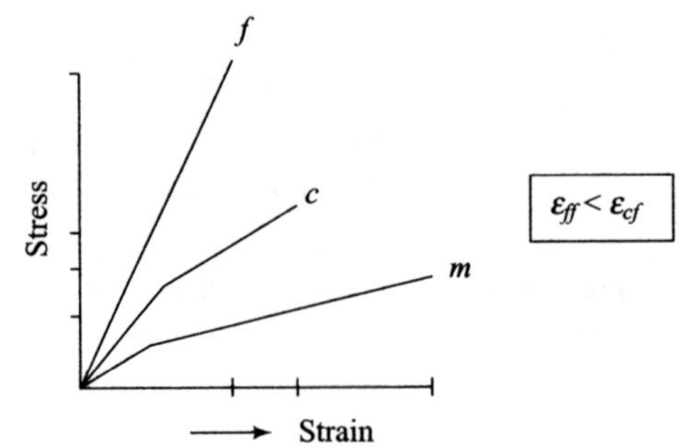

Figure 8.9 Fiber fails before composite fails

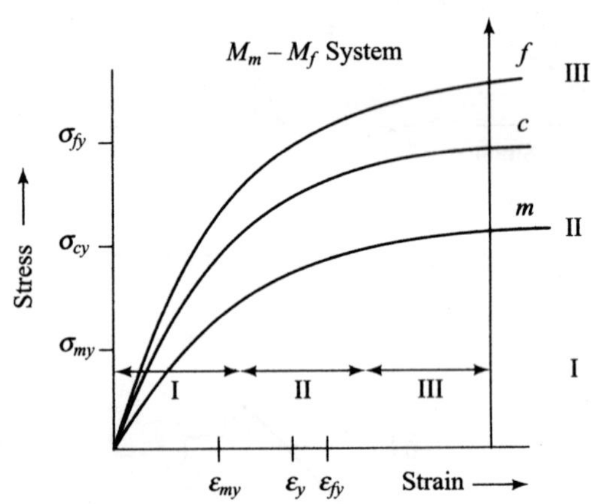

Figure 8.10 Stress against strain for a composite of metal matrix reinforced with a metal fiber

CHAPTER 9

MECHANICAL PROPERTIES

9.1 CREEP OF COMPOSITES

9.1.1 Introduction

Composites have better creep resistance than monoliths. Creep resistance increases when the volume fraction of reinforcement (V_f) increases. At higher stress (σ) creep resistance decreases as indicated by the higher steady state creep rate ($\dot{\varepsilon}_s$). As V_f increases the creep strain at which failure occurs also increases. Creep is a high temperature phenomenon. If T > ECT, it is high temperature. ECT (Equlcohesive Temperature) is the temperature at which the strength of grains is equal to the strength of grain boundaries. At temperatures above ECT, the grain boundaries are weaker than the grains. The ECT for a material depends on its melting point (T_m). The ECT is approximately equal to 0.5 T_m when all the temperatures are expressed in Kelvin.

Creep can occur under various circumstances, as explained herein.

Continuous fibres

The following cases are possible

1. The matrix creeps - fibres are elastic; stress is too low to cause damage to fibres.
2. The matrix creeps- fibres are elastic; stress is high enough to cause breakdown of fibres.
3. Both, matrix and fibres creep; but no damage to fibres due to very low stress.
4. Both matrix and fibres creep; and damage to fibres occurs due to high stress.

Short or discontinuous fibres

The following cases are possible:

1. The matrix creeps; but fibres deform elastically without creeping.
2. Both matrix and fibres creep

9.1.2 Creep of Continuous Fibre Reinforced Composites

Assessment of the creep behaviour of MMCs is a prerequisite for the application of such composites at high temperatures i.e ($T > 0.5\ T_m$). Although the stress to rupture or time to rupture factors do provide a comprehensive account of the creep behaviour of a material, they do not explain the mechanism of creep resistance. It is necessary that creep data is expressed as a creep strain-time curve. Refer figure 9.1 2124 aluminium matrix reinforced with 25% volume fraction (V_f) of SiC whisker, when tested for creep resistance (at 300°C and a stress of 50 MPa) produced a curve of creep strain vs time that exhibits three stages namely primary, secondary and tertiary. The majority of MMCs, behave in this manner because, generally the metal matrix creeps, but not the reinforcing fibre. This is especially true if the fibre is ceramic. Among the three stages the second stage that involves steady state creep is vital as it is of the longest duration. The slope of the creep curvey ($d\varepsilon/dt$) is constant for this stage and is equal to the steady state creep rate, $\dot{\varepsilon}_{s.}$ Creep resistant fibres are used to increase the resistance of a composite to creep. The composite Ag_m / W_f has a lower creep rate than the monolith, Ag alone, for a given temperature - stress combination. The creep rate of continuous fibre reinforced metal is more than that of discontinuous fibre reinforced metal which, in turn is more than that of a monolith. The steady state creep rate of a composite can be derived by applying the Rule Of Mixtures (ROM).

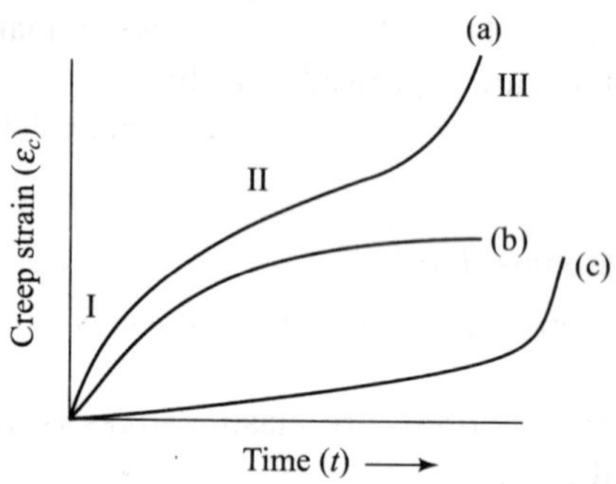

Figure 9.1 MMCs

$$\dot{\varepsilon}_{sc} = V_f\dot{\varepsilon}_{sf} + V_m\dot{\varepsilon}_{sm} \qquad (1)$$

where V_f and V_m are the volume fraction of the fibre and matrix $\dot{\varepsilon}_{sf}$ and $\dot{\varepsilon}_{sm}$ their steady state creep rates respectively.

The creep rate varies with different factors:

It increases with increase in the applied stress and decreases with increase in the fibre aspect ratio and fibre volume fraction.

The steady state creep rate $\dot{\varepsilon}_{sm}$ of a metal may obey either the Power law or the Exponential law. The former usually applies for low to intermediate stress level at high temperatures and the later for higher stresses at low temperatures.

The steady state creep rate of a composite, $\dot{\varepsilon}_{sc}$ can in some cases, be predicted from the steady state creep rates of the metal, $\dot{\varepsilon}_{sm}$, and that of the fibre $\dot{\varepsilon}_{sf}$ using appropriate model such as the Taya and Liholt model (that conforms to the exponential law). Equation 2 or the Street model (that conforms to the power law), equation 3 In such cases, the MMC may be tailor made for better creep resistance at specific application temperature.

$$\dot{\varepsilon}_{sm} = \exp(\sigma_m) \quad (2)$$

$$\dot{\varepsilon}_{sm} = A\,\sigma_{sm}{}^{n} \quad (3)$$

The basic equation that governs creep is,

$$\dot{\varepsilon} = A\,(\sigma/G)^n \exp(-\Delta Q/kT) \quad (4)$$

where

$\dot{\varepsilon}$ = creep rate, σ = applied stress G = shear modulus
ΔQ = activation energy, k = Boltzman constant
T = temperature in Kelvin.

A&n are constant

Assuming that creep of fibre and matrix obey power law we can write

Stress in fibre $\quad \sigma_f = \sigma_{fo}(\dot{\varepsilon}/\dot{\varepsilon}_{fo})^f \quad (5)$

Stress in matrix $\quad \sigma_m = \sigma_{mo}(\dot{\varepsilon}/\dot{\varepsilon}_{mo})^m \quad (6)$

Where σ is stress, $\dot{\varepsilon}$ is strain rate and superscripts (f&m) are empirical constants obtained from stress-strain relationships. The subscripts f, m and c respectively denote fibre, matrix and composite.

The composite creep rate is given by ROM. Assuming that both fibre and matrix creep at the same rate,

$$\sigma_c = \sigma_f V_f + \sigma_m V_m \quad (7)$$

$$= \sigma_{fo}(\dot{\varepsilon}/\dot{\varepsilon}_{fo})^f V_f + \sigma_{mo}(\dot{\varepsilon}/\dot{\varepsilon}_{mo})^m V_m \quad (8)$$

Alumina reinforced with 20% SiC whisker has a better creep resistance than monolith alumina. Normally, creep rates of fibres are higher than that of monolith ceramics. Fibres can be oxides or nonoxides. Nonoxide fibres with nonoxide matrices such as in the composites SiC_f/SiC_m and SiC_f/Si_3N_{4m} have better creep resistance though their oxidation resistance is poor.

Composite such as $C_f/Glass_m$, $SiC_f/Glass_m$, or Al_2O_{3f}/SiC_m which have either non-oxide fibre in oxide matrix or vice versa do not possess adequate oxidation resistance due to rapid oxygen permeation.

Creep resistance of laminates are generally better than that of monoliths. Creep resistance of unidirectional composite is superior to that of crossplied composite due to the fact that fibre oriented at 90° cannot contribute as much to creep resistance as fibre, oriented at 0° to the deforming force.

Composites of oxide fibre in oxide matrix show excellent resistance to oxidation. Thus, alumina fibre and mullite fibre are widely used with glass, glass-ceramic, alumina and mullite matrices. Two categories exist among such composites-coated oxide fibre with oxide matrix and uncoated oxide fibre with oxide matrix. With the former types, reactions take place between the coating and the fibre, as also between the matrix and the coating. Thermodynamics may be used to assess the stability of such composites. Kinetics also play an important role in determining the rates of such reactions.

Creep resistance of the composites, Al_2O_{3m}/ SiC_w (Refer figure 9.2) and mullite/SiC_w are better than that of Al_2O_3 or mullite as such. However, SiC_w and SiC_f get oxidised at temperatures above 1273K. Application of oxide matrix - oxide reinforcement composites are also limited by plastic flow at elevated temperatures under creep conditions. The strength of polycrystalline oxide fibres reduce substantially at temperatures above 1473K. At 1573K polycrystalline alumina or mullite suffers complete loss of strength. Single crystal (YAG) fibres can, however, withstand higher temperatures (above 1573 K). Oxides generally have low creep resistance. Thus $oxide_m/oxide_f$ composites have a shorter life at higher temperatures. For a creep rate of $10^{-8}s^{-1}$ at 1673 K no oxide material can provide resistance to creep (at a stress of 100 MPa and a failure strain of 1%). Consider a composite of polycrystalline alumina matrix with single crystal YAG fibre and another of single crystal alumina matrix and single crystal YAG fibre. The stress that can be withstood by these two composites are respectively 34 Mpa and 84 MPa at 1773 K for a creep rate of $10^{-8}s^{-1}$. Thus, a single crystal matrix offers better creep resistance than a polycrystalline matrix. Similarly single crystal fibres also offer better creep resistance than polycrystalline fibres. Therefore, for effective improvement of resistance to creep in a composite, the matrix and the fibre should both be of single crystal structure. In composites a weak interface promotes toughness by facilitating interface sliding while a strong interface is necessary for good creep resistance. Low volume of fibre / matrix bonding will increase the creep rate by a factor of 2 for ceramics reinforced with whiskers.

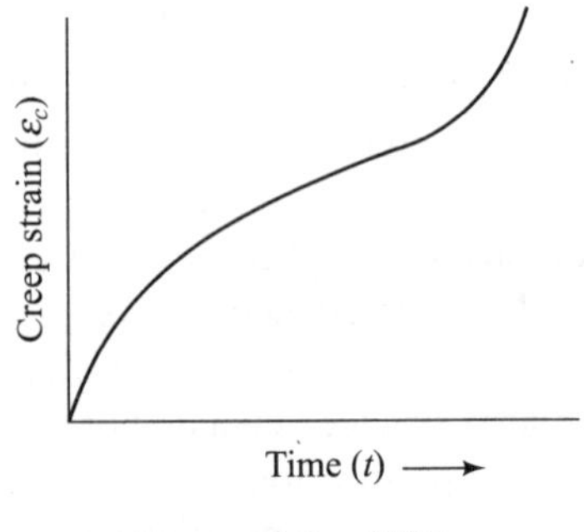

Figure 9.2 CMC_S

For continuous fibre composites assume an initial strain of ε_0, given by:

$$\varepsilon_0 = [\sigma / (V_f E_f + V_m E_m)] = \sigma_c / E_c = \text{initial strain} \quad (9)$$

This initial strain occurs instantaneously on loading a composite.

The applied stress, σ_c is given by,

$$\sigma_c = F_c / A_c \tag{10}$$

where, F_c is the force applied over an area A_c. The Young's modulus of the composite, E_c is given by

$$E_c = V_f E_f + V_m E_m \tag{11}$$

where V_f ***and*** E_f are the volume fraction and young's modulus of the fibre, and V_m and E_m those of the matrix

As creep occurs in the matrix, the applied stress is progressively transferred to the fibres. As fibres strain elastically, the stress in them increases. The limit is when the entire applied load is borne by the fibres,

This state is represented by the equation,

$$\varepsilon_\infty = (\sigma_c / V_f E_f) = \varepsilon_c = \varepsilon_f \tag{12}$$

This value is approached asymptotically. Since the rate of creep of the matrix falls off as the stress it carries decreases, and a steady state is never set up. The creep strain can never exceed a predetermined level, which is small assuming that the fibres do not break. If the matrix exhibits, power law creep in steady state, the creep rate of the matrix E_{sm} is given by,

$$\dot{\varepsilon}_{sm} = [d\varepsilon/ dt] = A\sigma_{sm}{}^{n} \tag{13}$$

and the creep rate of the composite is given by

$$\dot{\varepsilon}_{sc} = [A\sigma_{sm}{}^{n} / (V_m E_m + V_f E_f)] \, [(V_m E_m)(1 - \varepsilon/\varepsilon_\infty)^n] / (V_m)^n = [d\varepsilon/ dt]_{sc} \tag{14}$$

On integrating the above equation, strain as a function of time can be obtained.

9.1.3 Creep of Discontinuous Fibre Reinforced Composites

The composite deforms progressively and steady state creep occurs. The creep rate is strongly dependent on the creep characteristics of the matrix and the load carried by the fibres, which in turn depend on the aspect ratio of fibre as shown in Fig. 9.2. In GRP with $V_f = 0.5$ the stress in the matrix is reduced to one fifth when l/d changes from 1 to 10.

In PMCs, the matrix being viscoelastic and fibres are inert at usual service temperatures, it is the matrix that undergoes creep while fibres remain elastic.

When aluminium alloy reinforced with SiC_p for which (l/d = 1) and SiC_w for which (l/d = 5) are subject to creep loading it is found that the latter has lower ε_s than the former by a factor of two at any given stress level. Interface is also quite significant in determining creep resistance. When titanium alloy matrix with graphite coated SiC monofilament is subjected to creep at room temperature (303K) and at 773K, better creep resistance is shown in the former condition due to stronger cohesion. The decrease in creep resistance in the latter condition may be attributed to weaker cohesion.

The viscoelastic nature of the matrix in many fibre reinforced plastics (FRPs) causes the properties to be time and temperature dependent. Under constant stress, they exhibit creep, which will be more pronounced as the temperature increases. However, since fibres exhibit negligible creep, the time dependence of the properties of FRPs is much less than that of unreinforced matrix (monolith).

GRP with a volume fraction of 60% reinforcement when subjected to creep loading for 10^5 minutes fails at half (fifty percent) the tensile strength of the composite whereas under the same conditions, GRE fails at seventy percent of its tensile strength. Creep of laminates or discontinuous fibre composites is very rapid if matrix has a low creep resistance. Creep curve for PMCs are shown in Fig. 9.3. Creep resistance of typical composites are shown in Fig 9.4 (a) (b) and (c).

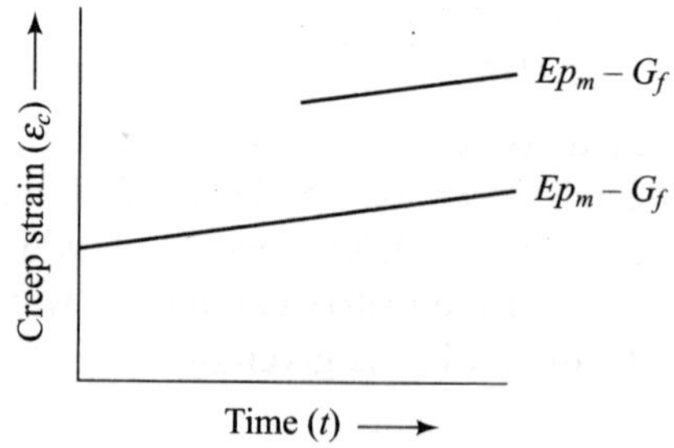

Figure 9.3 PMCs

9.2 FATIGUE OF COMPOSITES

9.2.1 Introduction

Materials fail at very low stresses under cyclic loading, which may be tension-compression, tension-tension or compression-compression as shown in Fig. 9.5. There are three stages in fatigue phenomenon namely, crack initiation, crack propagation and fracture. Fatigue cracks normally start at the surface in areas of inclusions, stress concentration etc. Here, the fatigue resistance of MMCs, PMCs and CMCs are elaborated. One cannot state that the fatigue resistance of composites is always better than that of their monoliths. However, by a proper choice of matrix and reinforcement, the fatigue resistance of composites are generally improved over that of their monoliths. Cyclic fatigue behaviour of composites is of great practical significance and is an important consideration in design engineering. In general moving parts such as automobile components are subjected to cyclic loading.

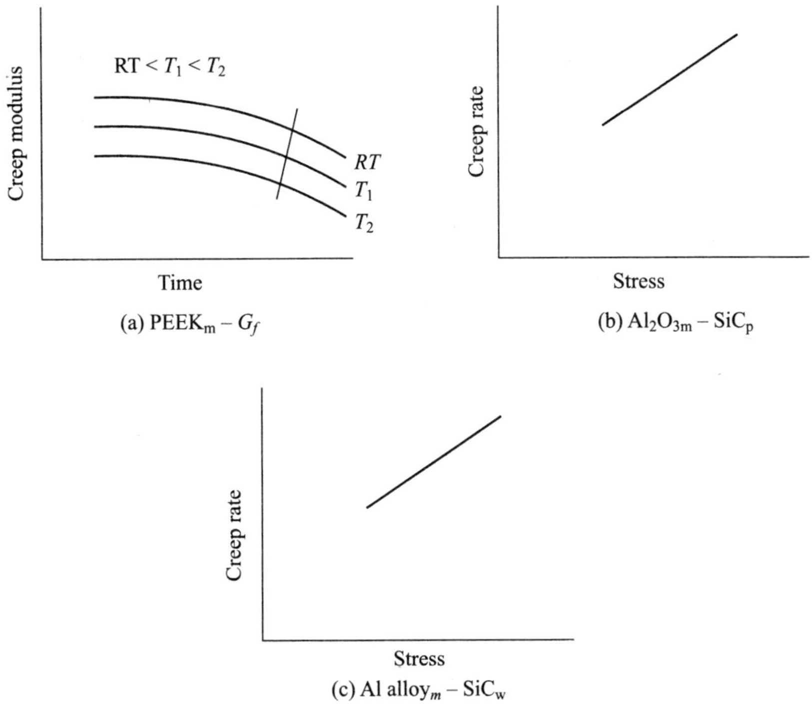

Figure 9.4 Creep resistance of composites

9.2.2 Fatigue of MMCs

There are three types of S vs log N (S-N) curves that represent fatigue of composites as shown in Fig. 9.6. Fatigue is caused by micromechanical damage due to (1) growth of matrix cracks in a direction perpendicular to fibre axis (Type A behaviour), (2) that of interfacial debonding (Type B behaviour) and (3) mixed effect of these (Type A & Type B).

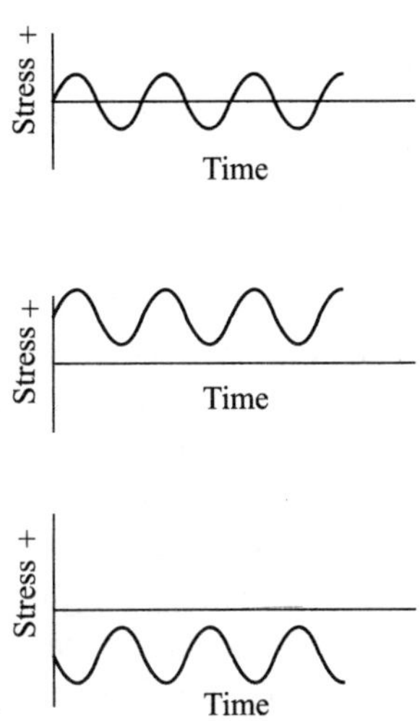

Figure 9.5 Types of loading in fatigue testing

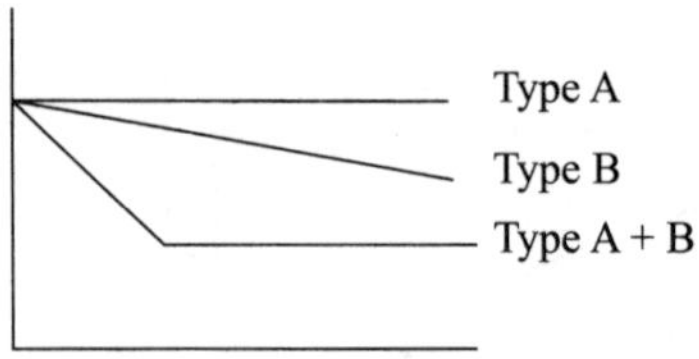

Figure 9.6 Fatigue behaviour of MMCs

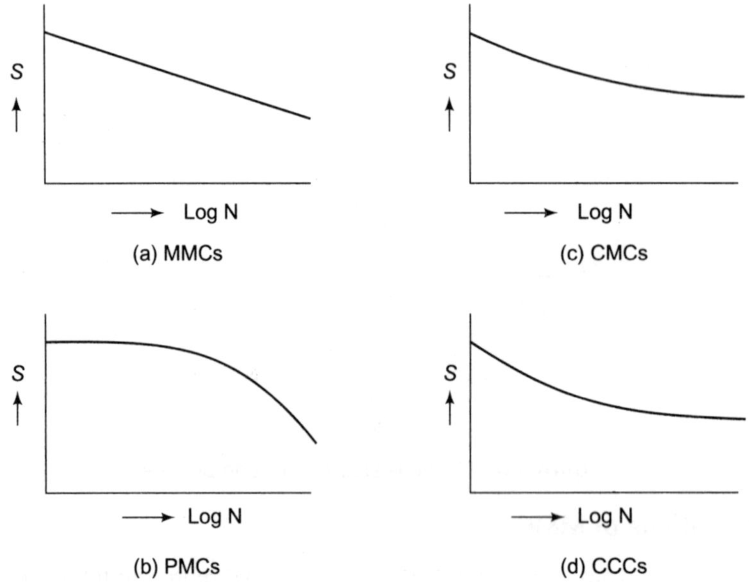

Figure 9.7 Plot of stress (S) vs number of cycles (N) showing fatigue resistance of composites

Type A behaviour of composites is similar to that of metals and alloys. Type B behaviour occurs when the fibres are perfect without weak points and debonding does not contribute to final fracture. In such a case the endurance limit will be the same as the static strength. In actual composites, majority of fibres have weak points (and therefore are not perfect) due to which they exhibit a mixed behaviour Type A & Type B.

Al_m/B_f composites have been studied and it is found that fibres break at weak points in the early cycles and thus act as nucleating agents for cracking in the matrix and for debonding. Such cracks grow progressively under fatigue loading finally leading to fracture and failure. The behaviour of 6061Al_m/B_f composite under fatigue loading conforms to the above model. The composite, (EZ41A) Mg $alloy_m/FP_f$ exhibits Type A behaviour at 300°K when matrix fibre interfacial bonding is strong (i.e. when no debonding has occured). Only transverse crack growth occurs and fibre/matrix strength ratio is relatively low. When the interface is weak i.e. when (debonding has occured) and fibre/matrix strength ratio is relatively high, Type B behaviour occurs. At 473°K the S-N curve shows a mixed behaviour due to the initiation of microcracks at the weak points of broken fibres leading to an interconnected network of the microcracks and matrix cracking.

One way of studying fatigue behaviour is by measuring the stiffness of the composite. As damage increases due to fatigue loading (more and more matrix cracking occurs) the Young's modulus of the composite decreases. As the stress increases the reduction in Young's modulus becomes more pronounced. This pattern of behaviour is shown by for Al_m/B_f composite. Fatigue loading does not necessarily lead to matrix cracking as is found in the case of Al_m/FP_f composite. Here, the interfacial bonding is strong and the fibre fractures extensively during fatigue loading and a crack propagates along a relatively straight plane. Multiple fibre fracturing occurs in Al_m/P_f composite. In 2124Alalloy$_m$/SiC$_w$ composite, numerous microcracks of varying lengths are formed at different cycles. The length (L) of these cracks obeys Weibull statistics. Crack length L increases with the number of cycles, N. The composite has better fatigue resistance than the monolith. A typical S-N curve is shown in 9.7 (a)

Particulate reinforced metals: The stress intensity factor ΔK threshold for MMCs is generally twice that for the monolith. MMCs have a shorter Paris Regime, and the exponent 'n' is higher than that for the monolith. Fracture toughness of MMCs is less than that of monolith. Due to these reasons, in the fatigue failure of MMCs, a single crack dominates; as in the case of monolith metal.

Fibre reinforced metals: The effects for particulate reinforced metals, wherein a single crack dominates as stated above are more pronounced in the case of fibre-reinforced metals.

9.2.3 Fatigue of PMCs

Fiber reinforced plasitcs are susceptible to fatigue. With thermoplastics as matrix materials there is the possibility of thermal softening failures at high stresses and/or high cyclic frequencies. The presence of fibres however, reduces the hysteresis heating effect and therefore, FRPs have a reduced tendency towards thermal softening failures. When thermal softening is contained by proper choice of conditions, normal fatigue takes place in the form of a progressive weakening of the material due to crack initiation and crack propagation. Plastics reinforced with Carbon or Boron fibres are stiffer than those reinforced with glass fibres. The former is less vulnerable to fatigue.

With discontinuous fibre GRPs, cracks tend to develop in the matrix and at the interface of the ends of fibres. If the matrix is a thermoset, cracks propagate through the matrix and destroy the integrity of the GRPs before the moulded article factures. Fatigue life depends directly on the aspect ratio, l/d of the fibres. As (l/d) increases, fatigue life increases.

At modest stress levels, debonding occurs after a small number of cycles. The build up of damage due to fatigue can be observed in translucent composites. The composite becomes opaque every time the load is applied. As the number of cycles increases the opacity becomes more pronounced and permanent. Subsequently, cracking starts and the resin matrix cracks will be visible. Still, the article is capable of bearing the applied load. Final separation occurs when intense localised damage has taken place. Thus there are three stages namely debonding, cracking and fracture in the course of fatigue failure. Usefulness may be limited up to the second stage itself due to aesthetic reasons or from safety consideration. Obviously, GRPs give a visual warning of the possible fatigue failure.

There is no definite fatigue limit. Often, it is necessary to design for a specific endurance. Usually a factor of safety of 3 to 4 is provided. Most fatigue data is for tensile loading with zero mean stress. To allow for other values for mean stress, the following empirical relationship is used.

$$\sigma_a = \sigma_f \ [1 - (\sigma_m / \sigma_c)] \tag{15}$$

where σ_f - fatigue endurance at N cycles

where σ_c - creep rupture strength at a time equivalent to N cycles.

Fatigue resistance of GRPs is lesser in other modes of loading such as torsion or bending. This is due to the inability of the composite to withstand shear stresses in sections of the matrix, which are unprotected by properly designed fibres. Reinforcement of plastics enhances the fatigue resistance, but in some cases, not in all. Moisture reduces the fatigue resistance of GRPs. Typical values of K for reinforced plastics are 5-50 $MNm^{-3/2}$. Plastics reinforced with carbon fibre show higher values.

The type and orientation distribution of fibre are significant. In composites, multiple cracking of matrix occurs unlike with monolith metals. Composite reinforced with stiff carbon or boron fibre show good fatigue resistance while those reinforced with glass of lower stiffness show poor fatigue resistance. Laminates have lower fatigue resistance and fail at lower loads than unidirectional composites. Chopped Stand Mat (CSM) and Dough Moulding Compound (DMC) have the poorest fatigue resistance. This is because the fibres in them are misaligned and have low aspect ratios. PMCs show low fatigue resistance in compression loading mainly due to the poor buckling resistance of fibres. Heat generated by high frequency cyclic loading cannot be conducted away by polymer matrices and fibres such as aramid or glass. During fatigue loading, temperature of the component raises leading to accelerated failure. A typical S-N curve is shown in Fig. 9.7 (b).

9.2.4 Fatigue of CMCs

For a long time it was thought that cyclic fatigue was of no particular significance for ceramics. Only as late as seventies did engineers realize the importance of cyclic fatigue studies on ceramics and ceramic matrix composites. These studies were rendered difficult by the problems inherently associated with all composites. The conventional approaches such as the S-N curves or linear elastic fracture mechanics (LEFM) are not quite suitable due to the fact that composites especially fibre reinforced composites, are heterogeneous and anisotropic. Therefore damage mechanisms in composites are very different from those encountered in conventional homogeneous or monolithic materials. In the S-N approach no distinction can be made between crack initiation and the crack propagation stages. In fatigue crack propagation tests the major problem is that one crack should markedly dominate as the failure pattern propagates. In composites, many modes of damage such as matrix cracking, debonding of matrix and reinforcement, crack deflection along interface; ply delamination and fracture of fibre are possible. Such a complex state of damage is to be characterized for drawing conclusions. One way of keeping track of the damage is to assess the progressive loss of stiffness. In an isotropic material, a single crack propagates in a direction perpendicular to the cyclic loading axis whereas in a fibre reinforced composite, a variety of subcritical damage mechanisms are active leading to a highly diffused damage zone. In a monolith, the single crack grows at a slow rate in the beginning and accelerates as the number of cycles increases. In a composite, especially in a laminate containing 0° and 90° plies, the damage proceeds at a fast rate in the beginning and then tapers off.

$Glass_m$-C_f composite showed no significant loss of strength when subjected to cyclic loading. For $LASI_m$-SiC_f, samples, the residual tensile strength and the elastic modulus were the same before and after fatigue. $LASII_m$-SiC_f, behaved similarly when subjected to cyclic loads below the proportional limit. For cyclic loads above the proportional limit, due to matrix micro cracking, the value of the elastic modulus of the test sample fell below the original value. Monolith LASI shows a linear, and LASII, a non-linear stress-strain relationship when subjected to simple tension. A typical curve is shown in Fig. 9.7 (c) for CMC and 9.7 (d) for CCC.

9.2.5 Fatigue of Hybrids and Laminates

As aspect ratio of fibres increases, the fatigue life also increases. As volume fraction V_f increases, fatigue life decreases. Long fibres can withstand higher stress for a given number of cycles than short fibres. Hybrids show longer fatigue lives than single fibre composites under identical conditions.

A hybrid composite of carbon fibre reinforced plastic (CFRP) and Graphite reinforced plastic (GRP) is better than either of these used alone. The strongest fibres do not always generate the most fatigue resistant composites.

TALREJA'S model suggests the existence of three stages in strain vs log N_f curves for PMC laminates.

Stage I: High stress level – Low life (N_f– cycles of fatigue). Fibre breakage occurs with or without interfacial debonding.

Stage II: Lower cyclic stress level. Statistical fibre breakage occurs but does not cause rapid composite destruction. Matrix cracking, interfacial shear failure etc. compete and influence overall damage. Slope of strain vs log N_f falls.

Stage III: Strain vs log N_f curve flattens and endurance limit is observed below some notional fatigue limit strain of the matrix. As long as working strains are below this limit no fatigue failure occurs.

In GRP laminates I stage does not occur; stage II starts directly.

For Kevlar fibre reinforced plastic (KFRP) laminates, stage I is of longer duration. The high stiffness of Kevlar fibre delays the setting in of stage II.

In CFRP working strains are so low that they do not exceed matrix fatigue strain limit. Stage I dominates.

9.2.6 Paris Law and Composites

Composites obey Paris law, which is represented by the equation given below:

$$(da / dN) = A(\Delta K)^p$$

Here, (da/dn) is the increase in crack length per cycle. ΔK is the stress intensity factor p is the paris exponent and A is a constant. The modulus ratio of fiber and matrix for composites are given is Table 9.1 below.

Table 9.1 Modulus Ratio

E_f/E_m	Composites
100	PMCs
10	MMCs
1	CMCs

9.3 FRACTURE TOUGHNESS

9.3.1 Introduction

Griffith proposed a theory of brittle fracture. According to this theory a crack will propagate when the decrease in elastic strain energy due to crack propagation is at least equal to the energy required to create the new crack surface.

$\sigma_{fr} = \left(\frac{2E\gamma_s}{\pi a}\right)$ in thin plates where the condition is plane stress and

$\sigma_{fr} = \left(\frac{2E\gamma_s}{(1-v^2)\,\pi a}\right)$ in thick plates where the condition is plane strain.

Orowan modified this equation so as to cover the criterion for materials, which undergo plastic deformation before fracture. The modified equation is,

$$\sigma_{fr} = \left(\frac{E\gamma_p}{a}\right)^{1/2}$$

$$\gamma_s = 1 - 2 \text{ J m}^{-2}$$

$$\gamma_p = 10^2 - 10^3 \text{ J m}^{-2}$$

Irwin modified Orowan's equation so that γ_p which is difficult of measure is replaced by the directly measurable G_c. The modified equation is,

$$\sigma_{fr} = \left(\frac{EG_c}{\pi a}\right)^{1/2}$$

where G_c is the critical crack extension force. It is given by

$$G_c = \frac{\pi a \sigma_{fr}^2}{E}$$

G can also be thought of as the strain energy release rate, i.e. the rate of transfer of energy from the elastic stress field of the cracked structure to the inelastic process of crack extension. The critical value of G which makes the crack propagate to fracture is termed G_c. It represents the fracture toughness of the material. The crack extension force G and the stress distribution around a flaw, which is represented by K, the stress intensity factor are related by the following equations:

$K^2 = GE$ for thin plates i.e. for plane stress condition and $K^2 = \dfrac{GE}{(1 - v^2)}$ for thick plates i.e. for plane strain condition.

K is more amenable to analytical determination than G, even though the latter has a more direct physical significance. Corresponding to G_c the critical crack extension force, there is a critical value K_c for the stress intensity factor.

9.3.2 Modes of Fracture

Interface in an ideal and a real composite are shown in Fig.9.8 (a) and (b). If the interface is strong failure of composite will be shown in 9.9 (a). If it is rather weak the failure will appear as shown in 9.9 (b). Cracking in composite occur as shown in 9.10 (b) and (c). Under different types of loading fracture will appear as shown in fig. 9.11 (b), (c) and (d). Delamination occurs as shown in fig.9.12(a) and (b).

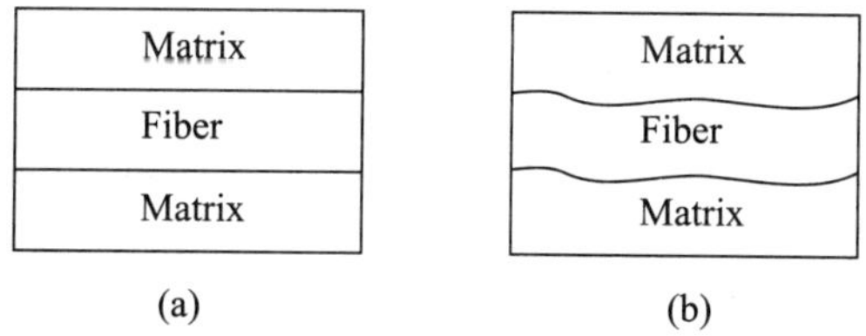

Figure 9.8 Interfaces (a) Ideal (b) Real

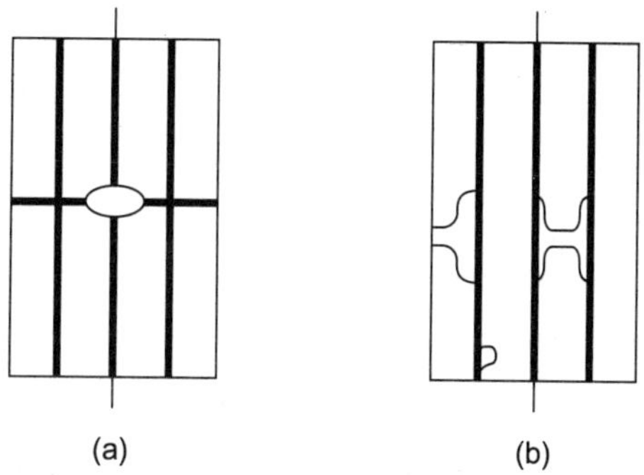

Figure 9.9 Interfaces (a) Strong (b) Weak

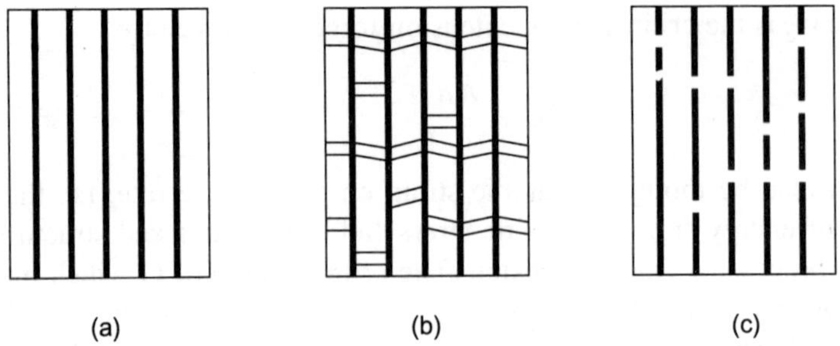

Figure 9.10 Cracking in composites

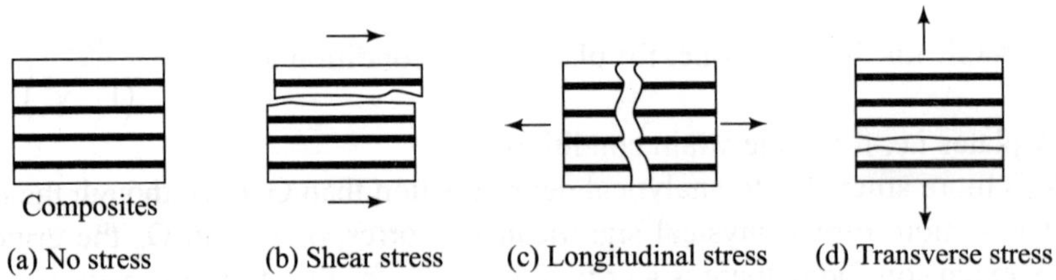

Figure 9.11 Fracture of composites under different types of loading

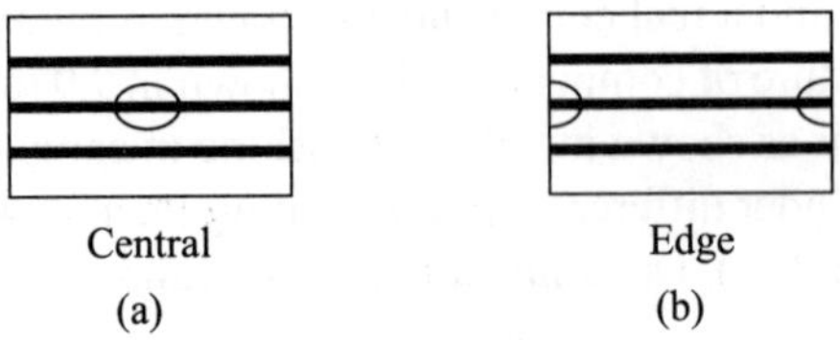

Figure 9.12 Types of delamination (a) central (b) edge

Fracture can occur in the following three modes as shown in 9.13(a), (b) and (c).

Mode I - Crack opening mode
Mode II - Forward shear mode
Mode III - Parallel shear mode

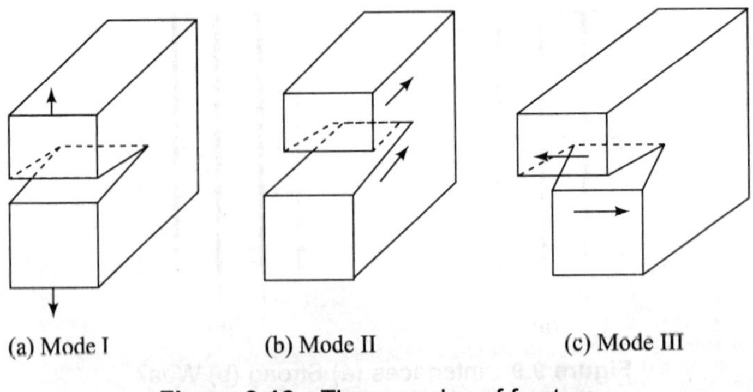

Figure 9.13 Three modes of fracture

Mode I represents the most severe condition. Critical value of stress intensity factor for this mode is termed as K_{Ic}. The plane strain condition is more severe than the plane stress condition. Plane strain fracture toughness is a valid material property, independent of specimen thickness especially for high strength materials. K_{Ic} is usually measured in the laboratories routinely. It is related to crack length as follows,

$$K_{Ic} = \sigma\sqrt{\pi a}$$

It is a true material property. The minimum thickness to achieve plane strain condition B, is given by the equation, $B = 2.5\,(K_{Ic} / \sigma_y)^2$

Where σ_y is 0.2% yield strength.

9.3.3 Determination of K_{Ic}

The standard specimens used for determining K_{Ic} are shown in Figs 9.14 (a), (b), (c) and (d).

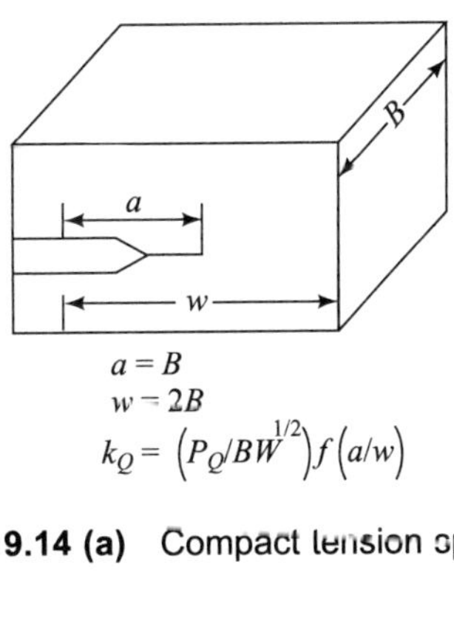

$$a = B$$
$$w = 2B$$
$$k_Q = \left(P_Q/BW^{1/2}\right) f\left(a/w\right)$$

Figure 9.14 (a) Compact tension specimen

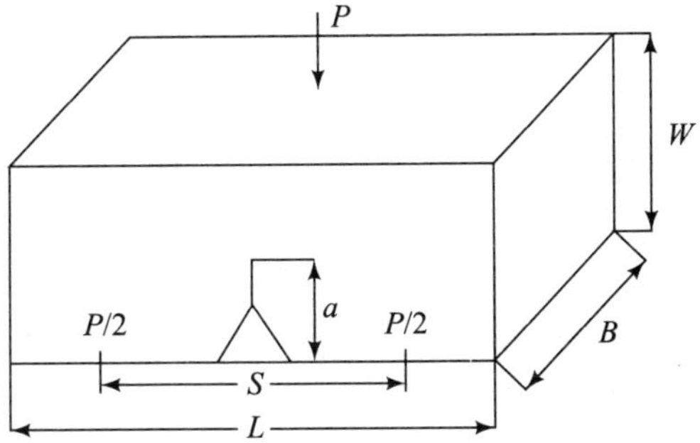

$$a = B$$
$$W = 2B$$
$$S = 8B$$

$$k_Q = \left(P_Q\, S/BW^{3/2}\right) f\left(a/w\right)$$

Figure 9.14 (b) Bend specimen

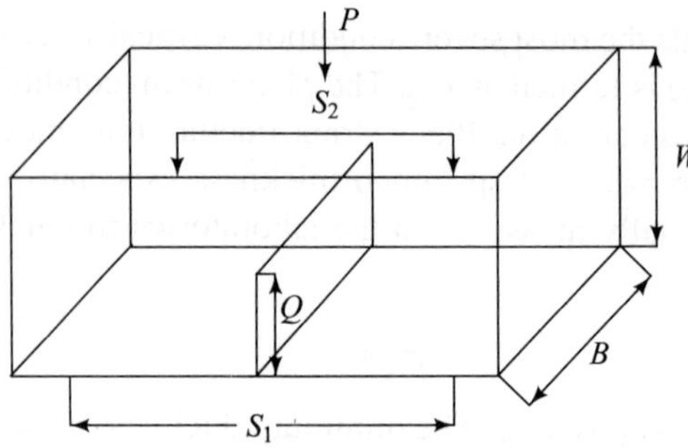

$$k_{IC} = \left[\frac{3\sqrt{Q}(S_1 - S_2)\, F_f n}{2BW^2} \right]$$

n = Calibration factor
F_f = Failure load

Figure 9.14 (c) Single edge notched beam specimen (SENB)

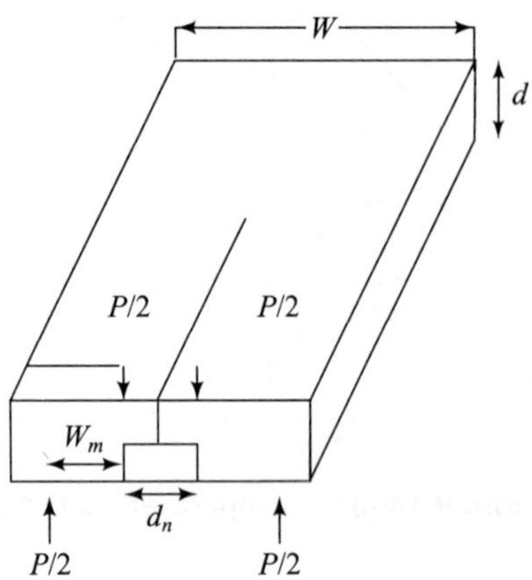

$$k_{IC} = PW_m \sqrt{\frac{3(1+V)}{Wd^3 d_n}}$$

Figure 9.14 (d) Double torsion specimen

The test specimen is subjected to load in a machine and the load against displacement across the open end of the notch are recorded to trace load vs displacement curves. There are three types of load vs displacement curves as shown in Figs 9.15(a), (b) and (c).

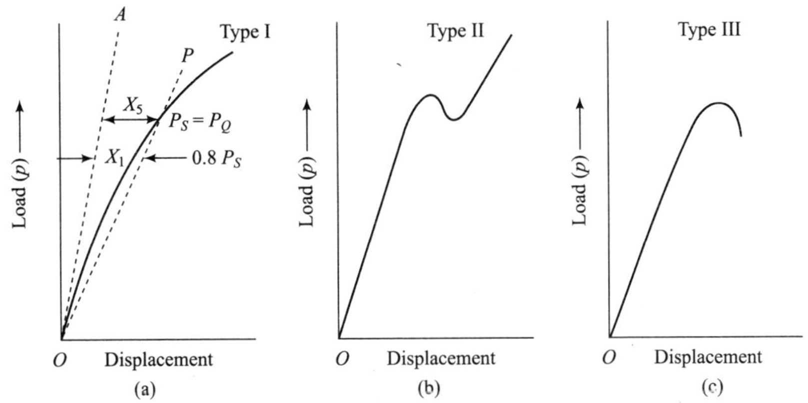

Figure 9.15 Load-displacement curves for determining fracture toughness

Type I is for ductile materials (plastic)
Type II is for intermediate materials
Type III is for brittle materials (elastic)

For each curve we have to determine a term P_Q. The procedure for this is described separately for each curve as follows.

9.3.3.1 *Type I*

Load Vs dispalcement curve rises continuously. Draw a tangent OA at origin O. Draw a secant line OP_s from the origin with a slope which is 5% less than that of the tangent OA. It cuts, the load displacement curve at P_s. Draw a horizontal line at a load of 0.8 P_s and measure the distance X_1 as shown in Fig. If X_1 exceeds ¼ X_s (X_s is at P_s) the material is too ductile to obtain a valid K_{Ic}. If $X_1 <$ X_s then $P_s = P_Q$.

9.3.3.2 *Type II*

There is a sharp drop in the load followed by an increase. The drop in load is due to unstable crack propagation. The procedure for determining P_q is the same as described for the Type I curve.

9.3.3.3 *Type III*

There is unstable crack propagation resulting in load drop due to complete failure. P_q is determined in the same way as for the other curves.

Once P_q is determined, another quantity K_Q has to be determined.

$$K_Q = \frac{P_Q}{BW^{1/2}}\left[f\left(\frac{a}{w}\right)\right] \quad \text{(for compact tension specimen)}$$

$$K_Q = \frac{P_Q\, S}{BW^{3/2}}\left[f\left(\frac{a}{w}\right)\right] \quad \text{(for bend specimen)}$$

In the above equations, a is the crack length. It is measured after fracture occurs. The value of B is given by $B = 2.5\,(K_Q / \sigma_y)^2$. If this is less than both the thickness of the specimen and the length of the crack in the specimen designate $K_Q = K_{Ic}$. Otherwise, repeat the test using a thicker specimen. The new specimen thickness is calculated using the equation $B = 2.5 (K_Q / \sigma_y)^2$ and putting herein the value of K_Q already determined.

9.3.4 Typical Values of K_{Ic}

K_{Ic} values of composites depend on percentage reinforcement as shown in figures 9.16 (a),(b) and (c). Fracture toughness of some of the composites is given in Table 9.2. Different standard specimens are used for MMCs, PMCs and CMCs.

Table 9.2 Toughness of Some Composites & Matrices

Material	K_{Ic} (Mpa $m^{1/2}$)
SiC	2 – 4
Alumina	1 -3.0
Alumina – SiC_w	3-5
Aluminium	100 – 350
Al – Al_2O_3	4 – 8
Soda Glass	0.7
Boro Silicate (Pyrex)	0.8
Epoxy	0.3 – 0.5
$Glass_f$ $Epoxy_m$	42 – 60
Aluminium Alloys	20 – 45
Aluminium Alloys + SiC_p	5 – 25
GFRP	20 – 60
Boro $silicate_m$ – C_f	-
Boro $silicate_m$ – SiC_f	50
$Glass_m$ – Al_2O_{3f}	7.0
ZrO_{2m} - SiC_w	11
CFRP	30 – 45
Titanium Alloys	50 – 100
Nickel Alloys	> 100
Copper Alloys	30 – 120
Steels	80 – 170
Cast Iorns	6 - 20
Glass – 25% SiC_w	2.1 – 3.4
Si_3N_4 – 30% SiC_w	7.5 – 10
Mullite – 20% SiC_w	4.6
Toughened Al_2O_3 - SiC_w	8.5 – 13.5
$MoSi_2$ – 20% SiC_w	8.2
ZrO_2	6.2
LAS	1.5
LAS - 50% SiC	17.0

9.3.5 Toughening Mechanisms in Composites

During crack propagation, several toughening mechanisms, could be operative in a composite. In a particular composite a particular mechanism will be dominant even though more than one mechanism may be possible. Toughening of a composite depends on (i) Size, shape and volume fraction of the reinforcement (ii) Interfacial bond between matrix and reinforcement (iii) Properties of the matrix and reinforcement and (iv) Phase transformations. For different composites different mechanism may be dominant and in some composites no one mechanism need be dominant. Following are some of the toughening mechanisms (i) Crack bowing (ii) Crack deflection (iii) Debonding (iv) Fibre pull out (v) Wake toughening or fibre bridging (vi) Microcrack toughening and (vii) Transformation toughening.

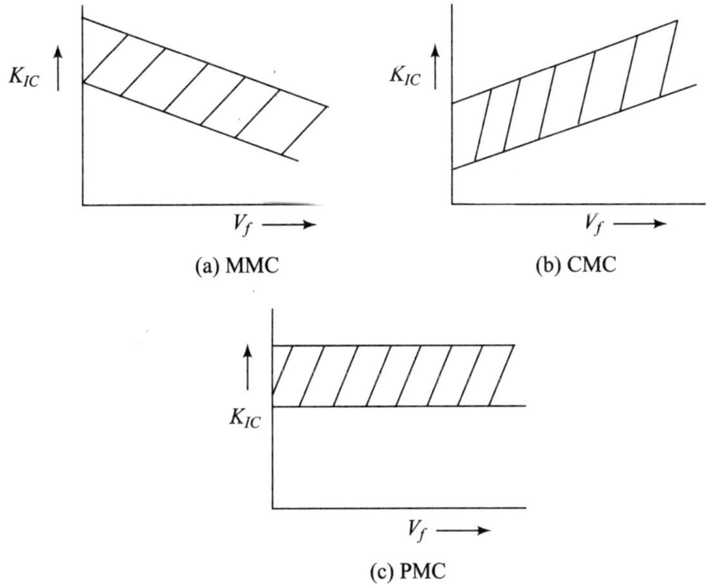

Figure 9.16 Variation of fracture toughness against volume fraction of reinforcement (schematic) (a) MMC (b) CMC (c) PMC

CHAPTER 10

TESTING AND INSPECTION

10.1 INTRODUCTION

New testing methods are required to evaluate the chemical, physical and mechanical properties of engineered structural composites since their properties are anisotropic. So also, different standard specimen shapes are required to evaluate the properties of reinforcement, matrix material, lamina and laminates.

10.2 REINFORCEMENT TESTING

The over all behaviour of a composite is greatly dependent upon the properties of the reinforcement. The physical and chemical properties of the reinforcement can be studied by conducting various tests.

10.2.1 Chemical Tests

Various chemical tests are commonly conducted on the reinforcement of composites. Some of these are described hereunder.

10.2.1.1 Surface analysis

X-ray photoelectron spectroscopy is chiefly used for surface analysis. It is also called Electron Spectroscopy for Chemical Analysis (ESCA). It is a node structure method for the chemical analysis of the surface. It gives total elemental analysis with the exception of hydrogen and helium. It is carried out at debts of (10-200 A°) from the surface. It should be done in vacuum. It is a reliable, easy to interpret, highly informative and sensitive quantitative test. However, being a sophisticated technique, it is quite expensive.

10.2.1.2 Sizing content analysis

This refers to determining the weight content of sizing in fibers.

Fibres with sizing are weighed; the sizing is removed and the fibres are weighed again. The weight content of sizing is then calculated. The sizing is usually removed by dissolving the fibres in a solution which does not affect the fibre material. Pyrolysis of sizing can also be done, without affecting fibre material.

10.2.2 Physical Tests

Physical tests are carried out for determining various physical constants of the reinforced material.

10.2.2.1 Density

Archimedes principle is applied to determine the densities of fibres.

10.2.2.2 Weight/unit length

A known length of fibre is weighed and the weight per unit length is calculated.

10.2.2.3 Filament diameter

This is relevant only for reinforcement with rounded cross section. The diameter of the fibre is measured directly by using microscopy.

It can also be calculated from the density of the fibre (filament) material, and the mass M of a known length L of the fibre.

$$\rho = \text{density} = \text{Mass/volume} = \text{mass}/\left(\pi D^2 L/4\right)$$

$$D = \sqrt{(4)\,(\text{mass})/\rho\pi L}$$

Mass is for a length of L of the fibre.

10.2.2.4 Electrical conductivity

Resistance for each of the known lengths of 25, 50, 75, 100, 125, 150 mm are measured using a resistance probe. The graph of resistance vs length is plotted. It will be a straightline, the slope of which gives the resistivity. Since resistivity depends on the diameter its value has to be mentioned.

10.2.2.5 Thermal expansion coefficient

The usual methods can be applied to measure the thermal expansion co-efficient of fibres.

10.2.3 Mechanical Tests

Among the various mechanical properties, tensile strength of fibre needs to be often determined.

10.2.3.1 Tensile strength of fibre

A filament is mounted on a slotted tab. The tabs are gripped so that the test specimen is aligned properly along the axis of the tensile testing machine and is tested at a constant speed Refer Fig.10.1(a) and (b). Testing is done at a constant strain rate till failure occurs. Cross sectional area of the filament can be measured using a planimeter or by optical gauges. Tensile strength and Young's modulus can be calculated from stress-strain curves.

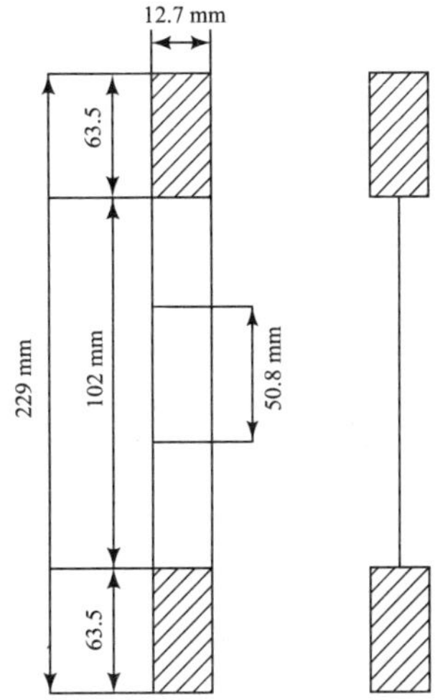

Figure 10.1 (a) (b) Tensile test specimen

10.3 MATRIX TESTING

Matrix materials can be polymers, metals, ceramics and intermetallics. These can be tested to determine chemical, physical and mechanical properties. The usual procedures that are well established and standardized are followed.

10.4 COMPOSITES TESTING

The characteristic properties of composites are studied by conducting destructive tests while those of particular components (finished composite products) are studied by conducting non-destructive tests on them.

10.4.1 Destructive Testing

These tests permanantely damage the specimen used for testing. Such tests cannot be conducted on finished products. The results of such tests serve as design guidelines.

10.4.1.1 Tensile test

The usual tensile test can be performed on unidirectional composites. Fibre glass tabs are placed on either side of straight sided specimens which, are gripped in the testing machine. Streamlined (reduction-section) specimen can also be used. However, it is very expensive to make such specimens and test them. Refer Figure 10.2. Flat specimen are shown in figure 10.3.

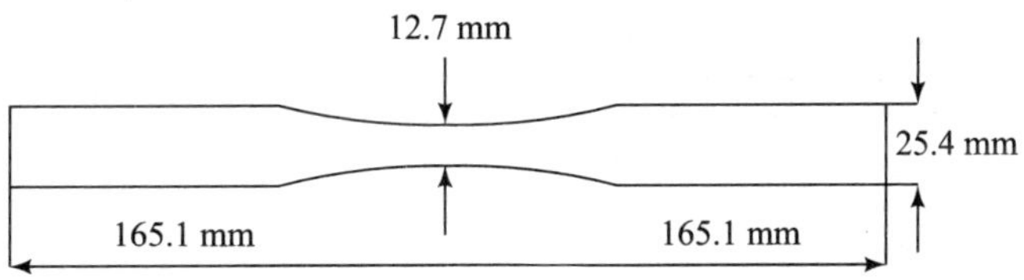

Figure 10.2 Stream lined specimen for tension test

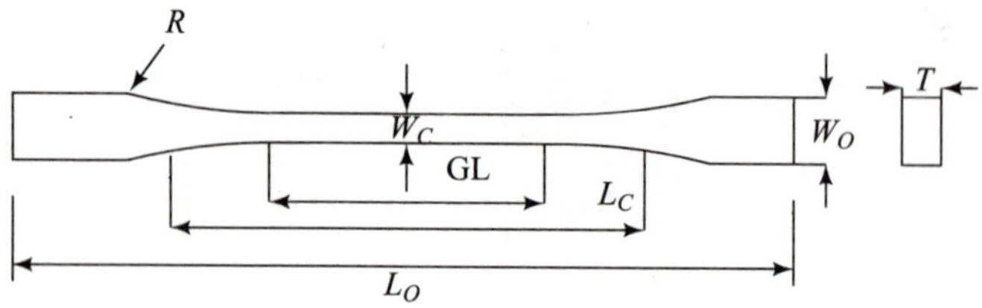

Figure 10.3 Sheet or Plate tensile specimen

T – Thickness = 3.2 mm
GL – Gauge length = 50 mm
L_o – Overall length = 165 mm
D – Distance between Grips = 115 mm
W_o – Overall width = 19 mm
L_c – Length of narrow section = 57 mm
W_c – Width of narrow section = 13 mm
R – Radius of fillet = 76 mm

10.4.1.2 Compression test

Here the test specimen should be prevented from buckling. One can use a cylindrical specimen for testing unidirectional composites. However, since this is liable to result in an underestimation of the strength, it is better to use conical wedges or pyramidal wedges for unidirectional composites.

10.4.1.3 Flexural test

This is an inexpensive test. It must however be noted that flexural strength is not an intrinsic property of a material. Three point bending or four point bending can be employed for this test. Span to depth ratio is kept at 32:1 for laminates. For aramid fibre composite it is kept at 16:1 for estimating strength and 60:1 for estimating modulus. Refer figure 10.4(a),(b) and (c).

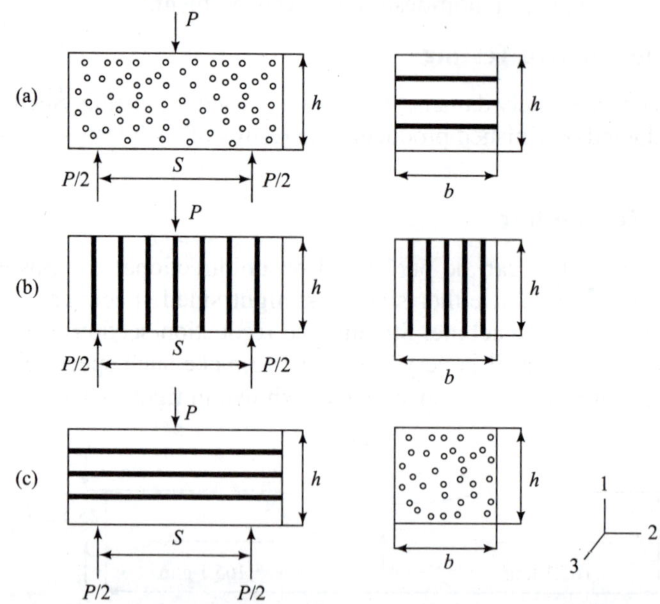

Figure 10.4 Three point bend test

10.4.1.4 Shear test

Iosipescu suggested this method for laminae, and short fibre reinforced plastics. A specimen with a double-V notch is used as shown in Fig 10.5. Both strength and modulus can be estimated. If the notch is carefully made, brittle materials can also be tested.

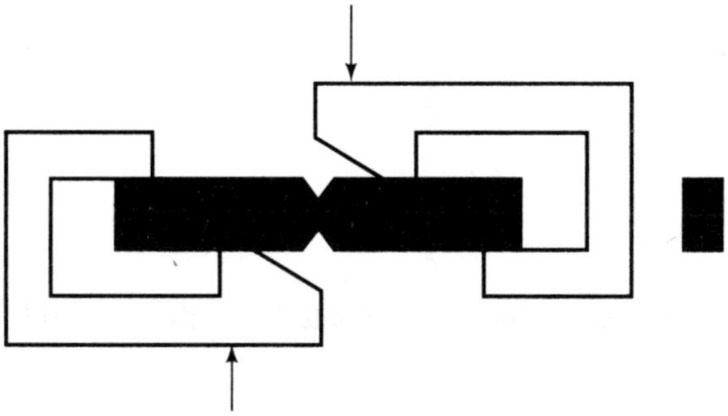

Figure 10.5 Specimen for Iosipescu shear test with holder for loading

Length (l) = 76 mm
Breadth (b) = 19.1 mm
Width (w) = 12.7 mm
Radius of notch = 1.3 mm
Angle of notch = 90°
Depth of notch = 4.3 mm

10.4.1.5 Interfacial shear strength test

Various tests have been developed to estimate the interfacial shear strength.

i. Fibre pullout test

It involves pulling a fibre embedded in a matrix material. Stress against strain can be plotted and the interfacial shear strength can be calculated. This test though simple in principle is difficult to execute. It is all the more so if the fibre is brittle. Refer figure 10.6

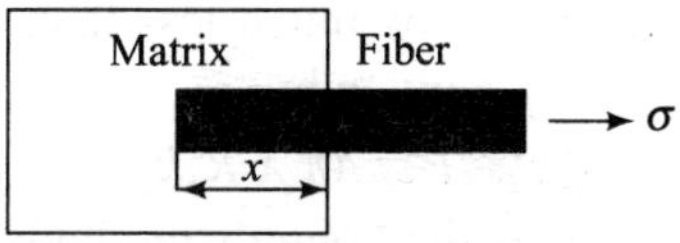

Figure 10.6 Specimen for fibre pull out test

ii. Compression test

A block of matrix material with a single short fibre in embedded position is subjected to compressive stress. The interfacial shear strength τ_I is given by,

$$\tau_I = 2.5\ \sigma_c$$

Where σ_c is the compressive stress at which debonding occurs. The debonding process can be observed accurately only if the matrix material is transparent. Refer figure 10.7

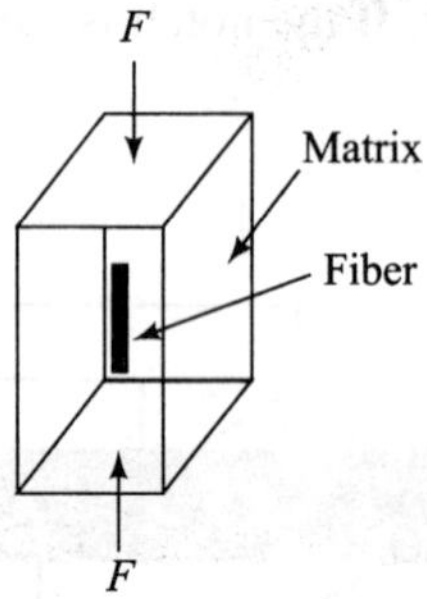

Figure 10.7 Specimen for interfacial shear strength

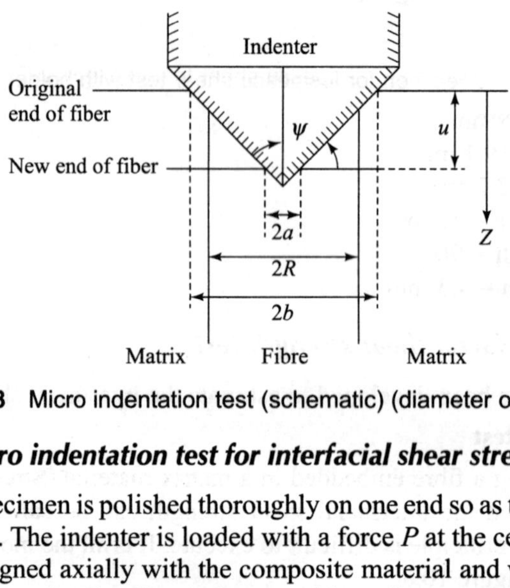

Figure 10.8 Micro indentation test (schematic) (diameter of fiber = 2R)

10.4.1.6 Micro indentation test for interfacial shear strength

A composite specimen is polished thoroughly on one end so as to be seen clearly in a microscope. The indenter is loaded with a force P at the centre of one fibre end which is aligned axially with the composite material and with the end surface normal to its axis. The fibre is depressed by a distance X, into the composite specimen. This distance should be much less than the specimen thickness. The interfaced shear strength τ_I is given by,

$$\tau_I = P^2 / 4\pi X R^3 E_f$$

where R is the radius and E_f is the Young's modulus of the fibre.

For a standard pyramid indentor,

$X = (b - a)$ cot 74° for a standard pyramid indentor.

Where, 2a is the indentation length on fibre and 2b - indentation length on matrix.

This test is very well suited for CMCs. Variation of τ_I with heat treatment parameters can be accurately measured, especially for CMCs, by this test. Refer figure 10.9.

10.4.1.7 *Interfacial tensile strength*

This parameter is often determined by the compression test

Compression test

A curved neck specimen with a single continuous fibre in embedded position is subjected to compressive stress and the debonding is observed. The interfacial tensile strength τ_I, is related to the stress in composite, τ_c, as follows, See Fig 10.9

Figure 10.9 Specimen for interfacial tensile strength

$$\tau_I = C\,\tau_c$$

C is a constant which depends on *v* and *E* of fibre and matrix.

10.4.1.8 *Impact test*

The schematic experimental set up for this test is shown in Fig 10.10. The specimen shape and dimensions are shown in Fig 10.11. There are two types of impact tests. (1) Instrumented impact test and (ii) Conventional impact test. In the former, one can know the stages in fracture process such as crack initiation and crack propagation as load is plotted against time. In the conventional method, the total energy absorbed before fracture occurs is obtained. The strain rate is of the order of 10^3 s^{-1} in impact tests. Triaxial state of stress exists in the sample near the notch. It is a severe test that subjects the specimen to a large stress and strain concentration. Velocity of the swinging pendulum is approximately 5 m s^{-1}.

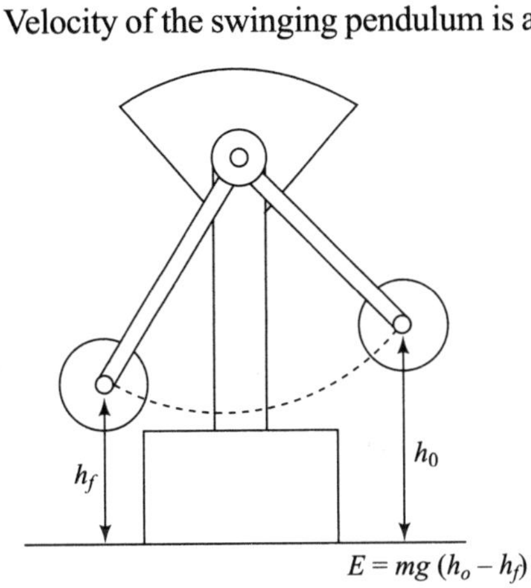

Figure 10.10 Impact testing machine

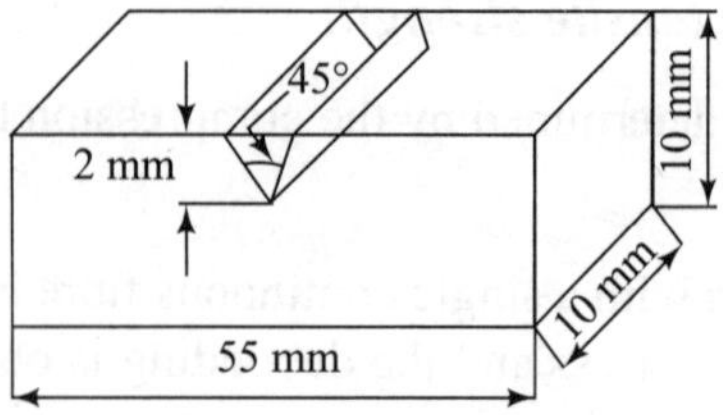

Figure 10.11 Charpy specimen

10.4.1.9 Creep test

The schematic experimental set up for this test is shown in Fig 10.12. The specimen used for testing creep in tension is of same standard as that used for simple uniaxial tension test. The specimen dimensions are shown in Figs.10.13 and 10.14. The applied stress should be much less than the yield stress of the material. Temperature employed should be greater than half the melting point of the material. Creep resistance can be found out in the constant stress mode or constant load mode. Normally there will be three stages in the creep strain vs time curve. These are, primary, secondary and tertiary creep

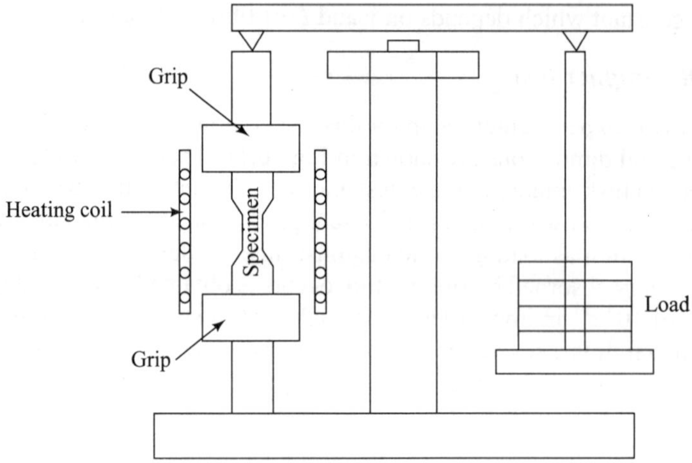

Figure 10.12 Creep testing setup

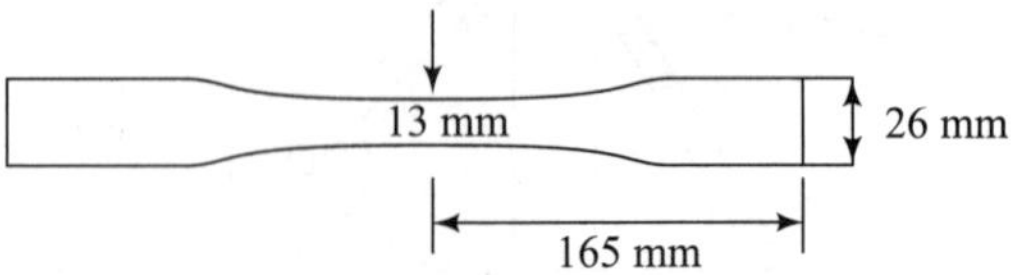

Figure 10.13 Streamlined creep test specimen (round)

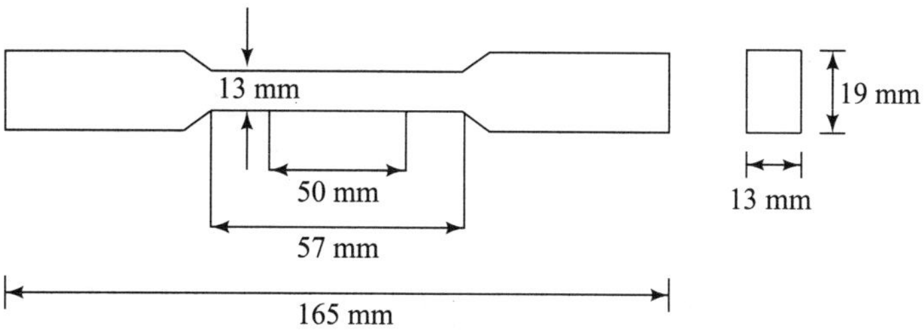

Figure 10.14 Sheet specimen for creep test

10.4.1.10 Fatigue test

The schematic experimental set up, and the specimen shape and dimensions for this test are shown in Figures 10.15(a), (b) and (c). The following three types of cyclic loading are possible,

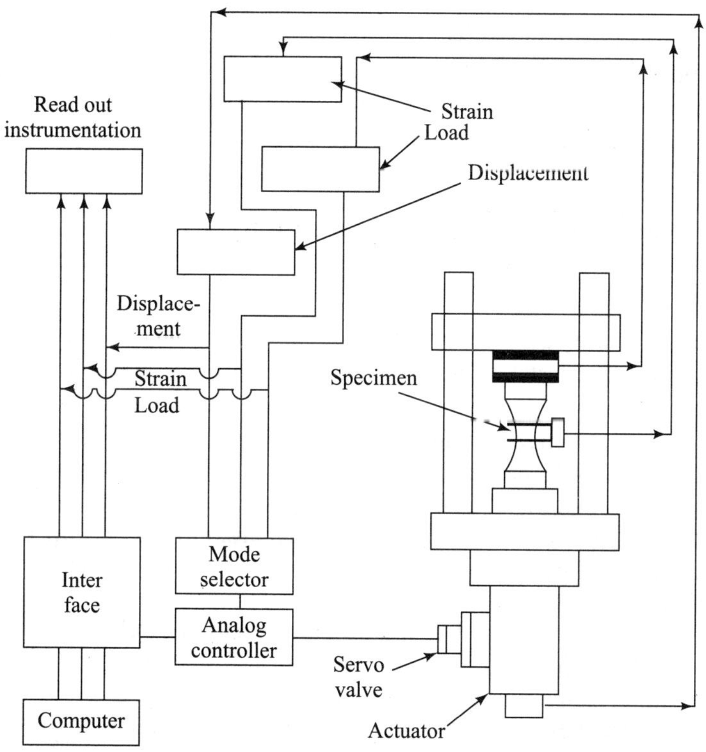

Figure 10.15 (a) Schematic diagram of a computer controlled closed-loop fatigue testing machine

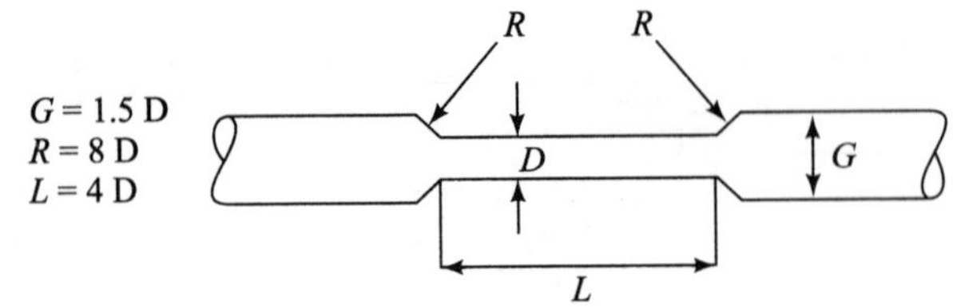

Figure 10.15 (b) Round fatique specimen

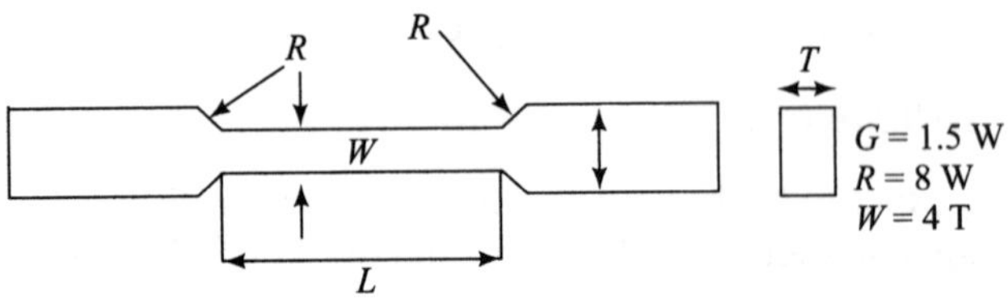

Figure 10.15 (c) Flat fatique specimen

Tension-Tension, Compression-Compression and Tension-Compression.

The applied stress should be much less than the yield point of the material. Stress vs log N curve is plotted. Fatigue limit or fatigue strength is obtained depending on whether stress is constant or decreases continuously with increasing cycles of applied load.

10.4.2 Non-destructive Testing/Evaluation

10.4.2.1 Introduction

To assess the quality and reliability of components made of composite materials, non-destructive tests/evaluation (NDT/NDE) are widely applied. Such tests are always done in combination with destructive testing. NDT/NDE cannot be a substitute for destructive testing but is complementary to it. There are several methods of NDT/NDE namely (1) Radiography, (2) Ultrasonic testing, (3) Liquid penetrant testing, (4) Magnetic particle testing and (5) Eddy current testing, called the "big five" methods and, (6) Thermal and infrared testing, (7) Microwave testing, (8) Acoustic emission testing, (9) Holographic testing and (10) Vibration methods.

These techniques are used to assess the severity as well as the existence, type, size and position of defects in composites. These are used to inspect the material without impairing the usefulness of the components. In the composite material industry these tests are mainly applied to PMCs and in research organizations, to PMCs, MMCs, CMCs, and IMCs.

10.4.2.2 Radiography

X-ray, γ-ray, electron and neutron, radiography techniques have been well developed and widely applied for metals. These techniques are based on the principle of differential absorption of *X*-rays, γ-rays etc. by various materials. *X*-ray radiography is usually applied to PMCs.

The general equation governing these tests is,

$$I = I_0 \exp(-\mu x)$$

Where,

I = Final intensity of the ray x = Thickness of the material

I_0 = Initial intensity of the ray

μ = Linear absorption coefficient of the material being tested.

μ/ρ is called the mass absorption coefficient. PMCs are usually constituted of hydrogen, carbon and other light elements with low μ/ρ values and therefore low energy X-rays are sufficient to inspect components made of PMCs as against those made of MMCs. A beam of 1 MeV X-rays can penetrate 500 mm of a polymer but only 150 mm of steel. Conventional low energy X-ray tubes, which are operated between 5-50 kV are sufficient for inspecting components of PMCs. Radiography is more suited to detect volume defects such as inclusions than narrow planar defects such as cracks. Refer figure 10.16.

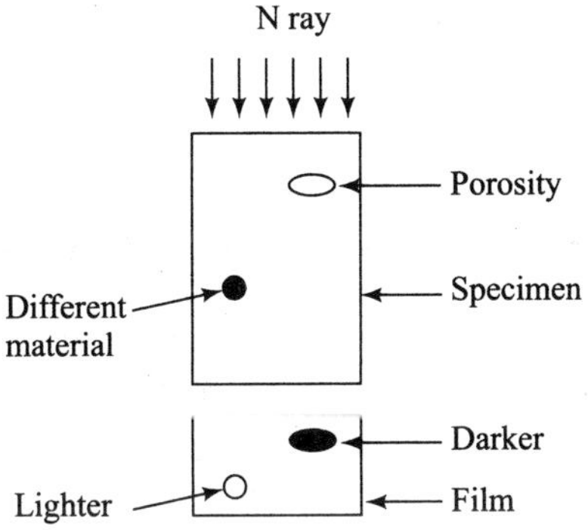

Figure 10.16 Radiography

10.4.2.3 Ultrasonics

The principle of acoustic impedance mismatch is availed for detecting defects by using ultrasonic sound waves of high frequency. These waves are inaudible to the human ear. They can be longitudinal waves, transverse waves or surface waves. Acoustic impedance of a material is given by, the equation,

$$Z = \rho C$$

where Z = acoustic impedance

ρ = density of material

C = wave velocity in the material.

If two materials with different Z values, Z_1 and Z_2 are coupled together and an ultrasonic beam is incident on the interface, the transmitted and the reflected wave intensities, (I_T and I_R) are given by the equations,

$$I_T = [4\,Z_1\,Z_2 / (Z_1 + Z_2)^2]\,(100) \text{ and}$$
$$I_R = [(Z_1 - Z_2)^2 / (Z_1 + Z_2)^2]\,(100)$$

so that

$$I_T + I_R = 100\%$$

The two different ultrasonic techniques widely used for testing components are, the pulse echo method and the transmission method. As shown in figures 10.17 and 18. A piezoelectric transducer is usually used to generate ultrasonic sound waves and a coupling medium such as water is used to inject the waves into the component.

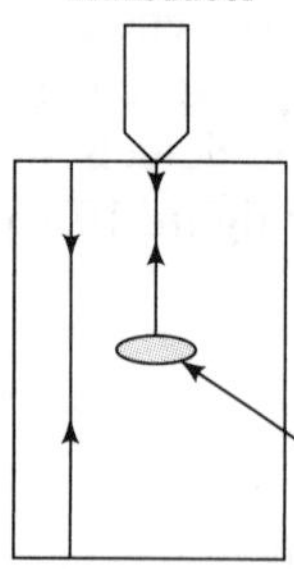

Figure10.17 Pulse echo method

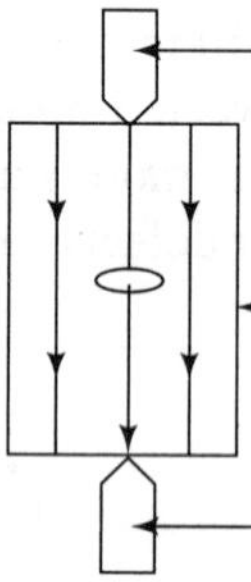

Figure10.18 Transmission method

Magnetic particle inspection cannot be used with PMCs because they are nonmagnetic. So also, the Eddy current technique cannot be used because PMCs are non conducting. Dye penetrant test is liable to cause problems if the absorbed dye is not completely removed after the testing. Ultrasonic testing and X-radiography are the most frequently used tests for PMCs.

For any nondestructive test, the resolution, accuracy, reliability and sensitivity of the technique employed are important. First the existence of a defect should be correctly established. Second, the results should be clear enough for one to draw conclusions based on them. Third, even small defects should be detected. Fourth independent defects should be identified correctly. They should not be misjudged as a single defect, which may lead to erroneous conclusions.

10.5 STATISTICAL ANALYSIS OF PROPERTIES

Composite materials are nonhomogeneous and anisotropic unlike conventional materials. Many of the properties of composite materials show a range of values and therefore it is imperative to assign statistical approximations to the property values which show inherent statistical variability. It is especially so for strength. If a property is given a basis value after analysis, it means that the probability of the property value exceeding this value is 95%. The survival probability is 99% for A basis and 90% for B basis.

Various schemes are available for statistically approximating values that show variations.

For a single batch, either Weibull distribution or normal distribution can be used. If both are found to be unsatisfactory, a nonparametric method can be applied.

For more than one batch, modified Lemmon method based on one way random effects ANOVA (Analysis of Variance) can be used. If batches can be pooled, the Anderson-Darling test can be applied.

10.5.1 Weibull Statistics

In 1951, Wallodi Weibull, a Swedish Engineer proposed the use of a mathematical function that could fit different shapes of survivor curves if the values of its three parameters are changed. There are two functions namely probability density function and reliability function or survivor curves. These two will have different shapes except in the special case where probability of failure is independent of age (i.e. there is a constant failure rate). Probability densities are proportional to the values of reliability functions. Weibull reliability function may be written in various ways. For instance,

$$S = \exp [-((X - X_u) / (X_0))^m]$$

m - Shape parameter
X_0 - Scale parameter
X_u - Location parameter

This statistical frequency distribution, as developed by Weibull, is an extreme value distribution and has been named after him Refer figure 10.19(a), (b) and (c). It is frequently used to analyse metallurgical/material properties and phenomena such as fatigue and fracture. It is used to characterise mechanical properties of brittle materials accurately and to replace the factor of safety concept with a more precise, strength reliability relation. $\sigma_{max} = (\sigma_f/0.5)$ where P = 0.5 but it is only $(\sigma_f/5)$ for P = 0.0001, for m = 70

where σ_{max} = maximum stress and

σ_f = fracture strength

when $X_u = 0$ and $X_0 = 1$

$$S = \exp [(-X)^m]$$

$$P(X) = 1 - \exp[(-X)^m]$$

$$p(x) = mX^{m-1} \exp[(-X)^m]$$

m is known as the Weibull modulus

p(X) - probability density function
P(X) - cumulative distribution function
$p(X) = dP(X) / dX$

X_0 and m can be obtained by using (i) Method of moments to solve the equations of P(X) and p(X), (ii) Graphically from P(X) and (iii) Maximum likelihood estimates. Values of weibull modulus for some materials are given in Table 10.1 below:

Table 10.1

Material	Weibull modulus 'm'
Steel	100
Carbon fibre	5 – 6
SiC, Si_3N_4 fibre	2 – 10
Al_2O_3 fibre	3 – 6
Mullite fibre	4
Chalk, Brick, Pottery and Cement	< 3

when $X = 0, \quad S = 1$

$X = X_0 \quad S = 1/e$

$S = 1 - P = \exp[-(X/X_0)^m]$

$\ln(1/S) = (X/X_0)^m$

and $\ln \ln(1/S) = m \ln(X/X_0)$

This is the equation of a straight line whose slope gives the value of 'm'. Smaller the value of 'm', greater is the variability in X. Therefore, the value of m tells us how rapidly the value of X falls as we approach X_0. Consider the case of the relation between strength and flaw size. With regard to weibulls distributions, the parameters have the values, $X = \sigma$, $X_0 = \sigma_0$, $X_u = 0$. Initially let there be a distribution of flaw with size of flaw at a particular value. Let us assume that due to improvement in processing the size of the flaw is reduced, but the distribution remains the same. i.e. 'm' remains same but mean strength increases. The new straight-line plot is parallel to the old one but is shifted to the right. Consider the case when the distribution changes but size remains constant. For the same mean strength, 'm' will be higher. These inferences are illustrated in Figs 10.20(a), (b) and (c). Larger specimens will have a higher probability of containing a larger defect, which in turn can cause lower strength i.e. larger the specimen, weaker it is likely to be. When the volume effect is taken into account, it can be shown that,

$$S = \exp[(-V/V_0)(\sigma/\sigma_0)^m]$$

This indicates that the survival probability depends on both, the volume subjected to the stress and the Weibull modulus 'm'.

Weibull statistics represents a versatile technique of statistical analysis often used in material sciences. Special Weibull probability paper is available to assist in the analysis. Though it is a relatively new technique it is rapidly developing in sophistication. It has been incorporated in FEM computer program for estimating stresses in brittle materials.

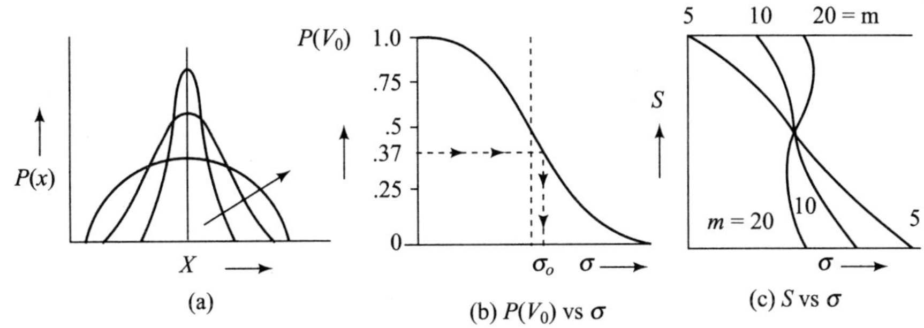

Figure 10.19 Weiball probability distribution function

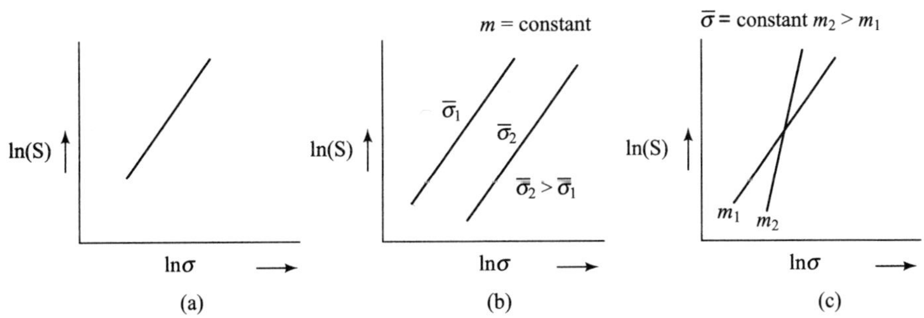

Figure 10.20 (a) (b) (c) Schematic variation of lnS with lnσ

$S = P(V_o)$ = Survival probability

V_o = Volume of sample

CHAPTER 11

DISPERSION AND PARTICULATE STRENGTHENED COMPOSITES

11.1 DISPERSION STRENGTHENED COMPOSITES

Such composites generally have a metal matrix in which fine oxide particles of the metal are disposed. The presence of the finely distributed oxide particles strengthens the matrix.

11.1.1 Introduction

Matrix is the major load-bearing constituent. The fine dispersions serve to impede the dislocation motion in the metal matrix. The matrix will be strengthened in proportion to the effectiveness of the dispersion as a barrier to the motion of dislocations. Dispersion strengthened composites are characterized by the following features.

Mean free path	0.3 – 0.01 μm
Inter particle separation	0.3 – 0.01 μm
Volume concentration of dispersoid	0.01 – 0.15 (1 – 15%)
Particle diameter	< 0.1 μm

11.1.2 Production Techniques

Following are some of the more common among the various techniques for producing dispersion-strengthened composites.

11.1.2.1 Surface oxidation

Sintered aluminium powder (Al matrix + Al_2O_3 particle) is prepared by this technique. First the raw materials are milled. This is followed by conventional operations such as compaction, sintering and hot extrusion. The strength of the composite increases as the amount of particle reinforcement increases.

11.1.2.2 Internal oxidation

Preferential oxidation of a solute element in a dilute solid solution is carried out. It results in the formation of fine oxide dispersion throughout the metal matrix.

The inward diffusion of oxygen must be faster than the outward diffusion of solute element. Cu_m - Al_2O_{3p} has been successfully produced by this route. Time, temperature, oxygen pressure and solute concentration are the process variables.

11.1.2.3 Reduction process

Starting material is a mixture of oxides of the matrix and the dispersoid. Initially thorough mixing is done. After this, the oxide matrix is selectively reduced.Subsequently consolidation techniques are employed. Thoria dispersed nickel is produced by this technique.

11.1.2.4 Mechanical alloying

It was first developed in 1970 for making dispersion-strengthened alloys. It involves a high energy ball milling process, for producing composites. There are three stages. During the first stage strong particle deformation and welding of particles take place. The dispersoid (for example, an oxide) is forced to cover the steadily increasing surface of the matrix (metal, for example) which embrittles continuously. In the second stage, the large lamellar like particles formed are broken and fractured to form new surfaces that pick up more of the oxide particles, reweld and form fine lamellae. In the third stage, an equilibrium is established between deformation and welding on one hand and fracturing on the other. Homogenisation proceeds and mechanical alloying stops. The dispersion phase can be any brittle phase (oxides, carbides, nitrides, intermetallics) while the matrix phase can be any ductile metal powder. An extensive range of composite materials can be produced by this technique.

11.1.3 Advantages and Applications

Dispersion strengthened composites maintain their yield strength upto a temperature of 0.8 T_m where T_m is the melting point in Kelvin of the matrix. They are insensitive to rise in temperature unlike precipitation hardening composite (in which softening and coarsening of precipitates occur at higher temperatures). The dispersion is a ceramic such as Al_2O_3, ThO_2, or SiO_2, which have a very high melting point as compared to that of the matrix metal, which could be aluminium, copper, nickel etc. Wherever elevated temperature strength, electrical and thermal conductivities are required, these composites can be used.

11.2 PARTICULATE STRENGTHENED COMPOSITES

These are composites in which the matrix is strengthened by an impregnant in the powdered form. A variety of materials can be used for the matrix and for the impregnation subject to the condition that they are mutually insoluable.

11.2.1 Introduction

Isostress condition is valid for these composites. Potential for strengthening is not as high as for fibre reinforced material. Yet these composites are widely used, due to substantial improvement in the other properties as compared to monoliths. They are characterized by; mean free path > 1μm, volume fraction of particle > 25% and particle diameter > 1μm.

11.2.2 Production Techniques

Finished components of particulate strengthened composites are manufactured by techniques, which generally involve powder metallurgical operations such as hot compaction.

11.2.2.1 Impregnation

A green compact is produced by cold or hot pressing a metal powder in a shaped die. The powder skeleton so prepared will be poor in mechanical strength but will have enough strength to withstand the handling during processing. It will contain 30 - 50% voids. This porous compact is placed in a furnace in contact with the impregnant and heated to above the melting point of the impregnant but below that of the metal constituting the porous skeleton. The impregnant and the metal are mutually insoluable. Ideal wetting should be ensured. The following combinations are popularly used. W/Cu, W/Ag, Mo/Cu, Mo/Ag and Ni/Ag. Unique combinations of properties are realized in these composites.

11.2.2.2 Solid state sintering

Two or more elemental materials that may or may not form an alloy during sintering are mixed, compacted and sintered. Only solid state sintering is effected. Extremely complex shapes of a variety of materials can be made by this technique. Final machining may be avoided in many cases. W/Cu, Ag/Gr, Fe/Cu and Ni/Ag are some of the composites usually produced by this technique.

11.2.2.3 Liquid state sintering

Binary or ternary mixture of ceramics and metals or metals and metals or ceramics and ceramics such as WC/CO, TiC/Ni, W/Ni/Cu, W/Ni/Fe, Fe/Cu, magnetite/wustite are pressed into compacts and then heated to a temperature such that a molten phase is formed by the reaction between the constituents. Sintering and densification take place by solution and precipitation followed by volume diffusion. The microstructure will consist of rounded refractory particles in a matrix phase.

11.2.3 Applications

Some of these composites are used as heat sinks, rocket nozzles, spark machining electrodes, electrical contacts, magnets, gears, bushings, weights, radiation shields, rock drilling bits, cutting tools and refractory bricks.

11.3 CERMETS

These are the earliest MMCs. These are combinations of ceramic and metals, produced by mixing the component powders, compaction and sintering. The phase dominating in terms of volume percentage is called the matrix and the other, the reinforcement or the dispersiod as in the conventional terminology. Examples of cements are Fe_m/Al_2O_{3p}, Mo_m/ZrO_{2p}, Al_m/V_3O_{8p} Al/U_3Al, Ag_m/CdO_p Ag_m/SnO_{2p}. These can be manufactured by conventional routes or by internal oxidation as for (Ag/Cd or Ag/Sn).

11.4 POLYMETS

Like CERMET, one can combine a polymer with a metal to get polymet which has properties not possessed by polymer alone. This can be achieved by mechanical alloying in a high energy ball or planetary mill. Polymer powder and metal powder can be combined in this fashion. Cryomilling can be done for better efficiency. For example by combining high density polyethylene with copper, one can get conducting polymet. By combining HDPE with iron ferromagnetic polymet can be obtained. By mechanical alloying one can get nano polymets also. A polymet is a PMC. Polymer can be polyethylene, polyaniline, PEEK, PMMA, HDPE etc. while metal can be Al, Sn, Cu, Fe, Ni etc. Inorganic polymers such as poly Silazzanc can also be used.

CHAPTER 12

RECENT DEVELOPMENTS

INTRODUCTION

Researchers and scientists of today are assiduously exploring the various possibilities offered by composites. The unprecedented advances in the fields of space science, medicine and defence have raised the demand for materials that possess high specific properties. Composites seem to be the only materials capable of meeting these requirements. A wide range of composites with interesting properties has come up.

12.1 SELF HEALING (REPAIRING) COMPOSITES

Damage in composites often begin as tiny cracks. As they grow, they progressively weaken the material until it breaks. To heal such tiny cracks automatically, capsules of glue are sprinkled throughout the fibre glass. The thicknesses of these capsules are equal to that of human hair. When a crack appears the capsules in its path break open spilling their contents and thereby sealing the cracks. The composite retains 75% of its original strength after the cracks have healed for 48 hours. Damage is thus, nipped in the bud. The concept has not yet been successfully commercialized. High temperatures render the healing time longer as the sealing agent is stopped from working. As of now, the time taken for curing is too long for such composites to be of any use in practical applications. Self-healing composites are part of the nascent field of smart materials. Objects such as spaceships, artificial joints and bridge supports which are inaccessible (difficult to reach while in service) are prime candidates for such self-healing composites. Computer circuit boards, which come out of the production line with small cracks, are also being targeted.

12.2 MOLECULAR COMPOSITES

Polyester is reinforced with a liquid crystalline polymer called Vectra (A plastic material five times stronger than steel and similar to Kevlar). By combining the polyester and the polymer at the right mixing speed and temperature, the Vectra forms fibrils that are embedded in the polyester and are attached to the polyester molecules. Polyester is ideal for making bonds with the liquid crystalline polymer.

12.3 MICRO COMPOSITES

In these composites, micro sized second phases such as particles, fibres and whiskers are dispersed at the grain boundaries of the matrix material to improve fracture toughness and strength. Matrix material is $Si_3 N_4$. Reinforcement can be SiC_w, SiC_p, TiC_p, TiN_p, TiB_{2p}, C_f, SiC_f.

12.4 NANO COMPOSITES

In these composites, nano sized particles are dispersed at the grain boundaries (intergranular) or within the grains (intragranular) of the matrix material or both or in another nano sized matrix material. Improvement in properties is very significant Machinability is improved. Superplasticity can be imparted to the matrix material. Si_3N_4 is the matrix. Nano sized particles of TiN or SiC are the reinforcements.

12.5 LEFT HANDED COMPOSITE MATERIALS

Refractive index of a material gives an idea about the bending of electromagnetic radiation (especially light). $\mu_{air} = 1.0$, $\mu_{water} = 1.3$, $\mu_{glass} = 1.5$ for light. A beam of light passing from air to water is deflected in one direction by a certain amount and is deflected by glass by a greater angle in the same direction. This bending helps curved glass surfaces to focus the incident light beam. Electromagnetic radiation traveling through ordinary materials is always deflected in the same direction giving those materials a positive index of refraction. A composite material made out of fibre glass and tiny copper wires bends microwaves in the opposite direction thus giving a negative index of refraction for the composite. The composite is called a "Metamaterial" ie. a composite fabricated from repeated elements specifically engineered to produce a desired electromagnetic behaviour. As on date this behaviour has not been observed with light but only with microwaves.

V.G. Veselago a Russian predicted such a behaviour in 1968. Colloquially such materials are called Left handed materials in contrast to conventional materials which are termed Right handed. This is to imply the reversal of various laws as applied to conventional materials. For instance, microwaves would emerge from the left handed composite in a direction opposite to that predicted by Snell's law. The Doppler effect is also reversed. As a microwave source approaches fixed object there is a shift to lower frequency and as the source recedes it shifts to higher frequencies, unlike sound waves for which the effect is opposite.

12.6 STIFFER THAN STIFF COMPOSITES

When ferro elastic vanadium oxide is incorporated in tin the stiffness of the composite is very high-even higher than that of tin embedded with the stiffest material, diamond.

Theoretically it is predicted that when a negative stiffness material is embedded in a conventional matrix material the overall stiffness can be increased by 40%. A material with negative stiffness exerts a force in the same direction as that of the applied force making it capable of accepting more deformation than a spring, which resists deformation and pushes back on squeezing.

The ferro elastic material has a negative stiffness and when it is incorporated and dispersed in a conventional material of positive stiffness, the resultant composite will have very high stiffness.

12.7 QUICK STEP PROCESS FOR PMCs

This process uses a unique fluid-filled balanced-pressure, floating mould technique combined with vibration through the fluids to produce advanced fibre and glass reinforced composite (PMCs) components. It is a fast fabrication method that does not use an autoclave. High quality of products can be achieved. It utilizes the thermal conductivity of fluids such as water to reduce the production time from 24 hours to 1hour for epoxy matrix composites. The apparatus required is less expensive and is not subjected to high pressure. No heavy construction is called for. The mould and the part are supported by fluid. Complex parts can be made faster and at lower costs. A fuselage or a boat hull can be made in a single process. By controlling the temperature of the fluid a greater precision can be achieved in curing. Energy consumption during production is less. Light vehicles fabricated with such composites will require lesser fuel in service.

12.8 BIOCOMPOSITES

The term biocomposites refers to those composites that can be employed in bioengineering, The primary aim in the development of biocomposites is that by varying the type and distribution on the reinforcing phases in the composite, it is possible to get a wide range of mechanical and biological properties and hence to optimize the structure and performance of biomedical devices and hence their reaction with the surrounding tissues. They are mainly used in the repair, reconstruction and replacement of human hard tissues. By controlling the volume fraction and local and global arrangement of the reinforcing phase the properties and design of a reinforcing phase can be varied and tailored to suit the mechanical and physiological conditions of the host tissues. Composite materials therefore offer a greater potential for structural compatibility than the homogeneous monolithic materials. Compoeite materials offer several other advantages over metal alloys and ceramics. For e.g., absence of corrosion and release of allergic metal ions such as Ni and Cr, higher fracture toughness and higher resistance against fatigue failure.

Biocomposites are classified based on their biodegradability, i.e., fully resorbable, partially resorbable and non-resorbable composites Resorbable biocomposites are made from those fibres and matrices both of which are fully absorbable in the body. These biocomposites are used for internal fracture fixation applications. Partialy resorbable biocomposites are fabricated using non absorbable reinforcing materials and absorbable matrix materials. They bave been investigated for applications such as bone replacement, bone cement and also internal fracture fixators. In a nonresorbable bio composite both the reinforcement phase and the matrix phase are non reasorbable in the body. The most advantageous use of nonresorbable biocomposites is in implants of the reconstructed or stabilized vertebral column.

The term biocomposites also refer to composites made of natural fibre. The use of fibres like flax, hemp, coir, jute or sisal in this industry so far is small since availablility of a durable semi finished product with constant quality is often a problem. Recent research and development have shown that thes aspects can be improved cosiderably. Knowing that natural fibres are cheap and have a better stiffness per weight than glass, which results in lighter components, interest in natural fibres is increasing. Secondly the environmental impact is smaller since the natural fiber can be thermally recycled and fibres come from a renewable resource. Finally natural fibres offer a possibility to developing countries to use their own natural resources in their composite processing industries. Importing synthetic fibers like glass with dollars can be avoided, and besides that, natural fibres if upgraded to a component or a semi-finished material like prepregs can become an export product for the country. Their moderate mechanical properties restrain the fibres from being used in high-tech applications, but for many reasons they can compete with glass fibres.

12.9 COMPLEX COMPOSITES

Complex composites provide multiple properties required by applications in power generation, construction, aerospace, packaging and other industries by combining two or more different materials, only where needed in a structure, part or package. Complex composites are custom designed to meet the requirements of particular applications and are combination of (1) Two or more different substrates (2) Two or more different resin systems or (3) A conventional composite material and a non composite material. Examples are (a) Steel pipes mounted on the outerside with a thermosetting plastic which are used in corrosion resistant applications such as for carrying water and chemicals, (b) Copper applied on a thermosetting plastic and latter machined to suitable geometries to generate the required magnetic fields, (c) A paper besed or cotton based core with a fibre glass besed outer material to give stiff panel member, (d) Lighter foam or honeycomb core material with a fiberglass outer covering material and (e) Thermosetting plastic inner core with a thermoplastic films as outer covering materials to get a combination of stiffness and flexibility. When the requirements of your applications are technically difficult or impossible to meet with conventional material you should opt for complex composities. It is economical and gives the required contradicting properties.

CHAPTER **13**

NUMERICAL PROBLEMS

(1) Compare the energy absorption capabilities of composites produced using carbon fibres, aramid fibres and glass fibres. Use the following data:

Fibre	σ (Gpa)	E (GPa)	$V = \sigma^2/2E$ (GPa)
C (HS)	2.9	230	1.83×10^{-2}
C (HM)	2.2	380	0.64×10^{-2}
Kevlar	3.0	130	3.46×10^{-2}
E-Glass	2.0	80	2.5×10^{-2}
S-Glass	3.3	91	6.0×10^{-2}

(2) A hybrid composite material is made up of 20% HS carbon fibres by weight and 30% E-glass fibres by weight in an epoxy matrix. Find the density of the composite.

		ρ (Kg/m^3)
V_{fc}	= 0.181	1800
V_{fg}	= 0.192	2540
V_{mep}	= 0.627	1300

(3) What weight of carbon fibres (ρ = 1800 kg/m^3) must be added to 1 kg of epoxy (ρ = 250 kg/m^3) to produce a composite with a density of 1600 kg/m^3.

$W_f = (\rho_f/\rho_c)V_f$, $W_m = (\rho_m/\rho_c)V_m$

$W_m : W_f = 0.284 : 0.716$

(4) A unidirectional glass fibre/epoxy composite has a fibre volume fraction of 60%. Given the data below, calculate the density, modulus and thermal conductivity of the composite in the fibre direction.

	ρ (kg/m^3)	E (GPa)	K (W/m.K)
Epoxy	1250	6.01	0.25
Glass fibre	2540	80.0	1.05

$\rho_c = \rho_f V_f + \rho_m V_m$

$E_c = E_f V_f + E_m V_m$

$K_c = K_f V_f + K_m V_m$

(5) In a unidirectional Kevlar/epoxy composite (E_f/E_m), the modulus ratio is 20 and the epoxy occupies 60% of the volume. Calculate the modulus of the composite and the stresses in the fibre and the matrix when a stress of 50 MN/m^3 is applied to the composite. The modulus of the epoxy is 6 GPa.

$$E_c = E_f V_f + E_m (1 - V_f)$$

$$\sigma_c = \sigma_f V_f + \sigma_m (1 - V_f)$$

$$\sigma_c/E_c = \sigma_m/E_m = \sigma_f/E_f \text{ (i.e., } \varepsilon_c = \varepsilon_m = \varepsilon_f)$$

(6) A reinforced plastic sheet is to be made from a matrix with a tensile strength of 60 MN/m^2 and continuous glass fibre with a modulus of 76 GPa. If the resin ratio by volume is 70% and the modular ratio of the composite is 25, estimate the tensile strength and modulus of the composite.

$$\varepsilon_f = \varepsilon_c$$

$$\sigma_f/E_f = \sigma_c/E_c$$

$$\sigma_f/\sigma_c = E_f/E_c$$

$$F_f A_c/F_c A_f = E_f/E_f V_f + E_m V_m$$

$$F_f/F_c = (E_f/E_m) / [(E_f/E_m) + (V_m/V_f)] = 40/(40 + 1)$$

$$F_f = 97.6\% F_c$$

(7) A matrix plastic material (T.S = 60 MPa) is reinforced with randomly oriented glass fibre (T.S. = 76 GPa, V_f = 30%) which is discontinuous. What is the tensile strength of the composite? The fibres are 15 mm long, have an aspect ratio of 1000 and produce a reinforcement efficiency of 0.25. The fibre strength is 2 GPa and the shear strength of the interface is 4 MPa.

(8) A polyester matrix is reinforced with continuous glass fibres. A 15 mm wide beam made from this material is to be simply supported over 300 mm length and have a point load at midspan. For a fixed beam weight of 90 g/m, investigate how the stiffness of the beam changes with the volume fraction of glass and state the optimum V_f ($\rho_f = 2560$ kg/m^3, $\rho_m = 1210$ kg/m^3, $E_f = 76$ GPa, $E_m = 3$ GPa).

$$\delta = PL^3/48EI$$

$$I = bd^3/12$$

$$P/\delta = 48EI/L^3 = 4Ebd^3/ L^3$$

$$E = Ec = E_f V_f + E_m (1 - V_f)$$

$$W_{Beam} = W_f + W_m$$

$$W_{Beam} = (\rho_f V_f + \rho_m V_m) (db)$$

$$d = W_{Beam}/b(\rho_f V_f + \rho_m (1 - V_f))$$

Stiffness $= (4/ L^3 b^2) (E_c)(W \times 10^3/\rho_f V_f + \rho_m V_m)$

Maximum stiffness occurs at $V_f = 0.37$

$$W_f = (\rho_f V_f)/(\rho_c V_c) = w_f/ w_c = (\rho_f/\rho_c).\ V_f$$
$v_f =$ Volume of fibre $w_f =$ Weight of fibre

$v_c =$ Volume of composite $w_c =$ Weight of composite

$W_f =$ Weight fraction of fibre

$V_f =$ Volume fraction of fibre

(9) The density of a composite made from unidirectional glass fibres in an epoxy matrix is 1950 kg/m^3. If the densities of the glass and epoxy are known to be 2540 kg/m^3 and 1300 kg/m^3, respectively, calculate the weight fraction of fibres in the composite.

$$\rho_c = \rho_f V_f + \rho_m V_m$$
$$V_f = 0.52$$
$$W_f = (\rho_f/\rho c)\ V_f$$

(10) A PEEK matrix is to be reinforced with 30% by volume of unidirectional carbon fibres and the properties of the individual materials are given below. Calculate the density, modulus and the strength of the composite in the fibre direction.

	ρ (kg/m^3)	T.S (GPa)	E (GPa)
PEEK	1300	0.058	3.8
Carbon fibre	1800	2.1	400

$$\rho_c = \rho_f V_f + \rho_m V_m$$
$$E_{cL} = E_f V_f + E_m V_m$$
$$\sigma_{cu} = \sigma_{fu} V_f + \sigma_{mu} V_m$$

(11) Calculate the fraction of the applied load which will be taken by the fibres in the composite in the above problem.

$$F_f = \sigma_f A_f = E_f\, \varepsilon_f\, V_f\, V_c$$
$V_f = v_f / v_c =$ Volume of fibre/ Volume of composite

$v_f = A_f$ for uniform cross section

$$A_f = V_f.\ v_c$$
$$F_{cl} = E_f\, \varepsilon_f\, V_f\, v_c + E_m\, \varepsilon_m\, V_m\, v_c$$
$$\sigma_{cl} = \sigma_f\, V_f + \sigma_m\, V_m$$
$$\sigma_{cl}/\sigma_f = V_f + (\sigma_m/\sigma_f)\,(1 - V_f)$$
$$\sigma_m/\sigma_f = E_m \varepsilon_m/\ E_f \varepsilon_f$$
$$\sigma_{cl}\, \sigma_f = V_f + (E_m\, E_f)\,(1 - V_f)$$
$$\sigma_f/\sigma_{cl} = 3.26$$

(12) For the PEEK/ Carbon fibre composite referred to above in problem 10, calculate V_{min} & V_{cri} UTS of PEEK is 62MPa.

$$V_{cri} = (\sigma_{mu} - \sigma_m')/(\sigma_{fu} + \sigma_{mu} - \sigma_m')$$
$$V_{min} = (\sigma_{mu} - \sigma_m')/(\sigma_{fu} - \sigma_m')$$

(13) Calculate the transverse modulus of the PEEK/ carbon fibre referred to in problem 10 considering poisson's ratio as $v_m = 0.36$.

(14) A laminate is 300 mm wide by 10 mm thick and is made up of 3 layers consisting of unidirectional fibres aligned in loading direction, the top layer with 60% volume of glass fibres, middle layer with 33% volume of carbon fibres and the bottom layer with 40% volume of aramid fibre. Matrix is epoxy in all the three layers. Calculate the load carrying capacity of the laminate. After one layer of the laminate becomes ineffective, what would be the load carrying capacity?

	UTS (GPa)	E (GPa)	b x t
Epoxy	0.07	6	
Glass fibre	2.1	73	300 x 3
Carbon fibre	2.8	240	300 x 4
Kevlar fibre	3.0	130	300 x 3

Load is applied perpendicular to (b x t) area.

Glass/ Epoxy $E_{G/Ep} = 46.2$ GPa $\sigma_{G/Ep} = 1.29$ GPa

Carbon/Epoxy $E_{C/Ep} = 83.2$ GPa $\sigma_{C/Ep} = 0.97$ GPa

Kevlar/Epoxy $E_{K/Ep} = 55.6$ GPa $\sigma_{K/Ep} = 1.24$ GPa

$$\sigma_{G/Ep} = E_{G/Ep}\left(\frac{\sigma_{C/Ep}}{E_{C/Ep}}\right) = E_{G/Ep}\left(\frac{\sigma_{K/Ep}}{E_{K/Ep}}\right)$$

$$\sigma_{C/Ep} = E_{C/Ep}\left(\frac{\sigma_{G/Ep}}{E_{G/Ep}}\right) = E_{C/Ep}\left(\frac{\sigma_{K/Ep}}{E_{K/Ep}}\right)$$

$$\sigma_{K/Ep} = E_{K/Ep}\left(\frac{\sigma_{G/Ep}}{E_{G/Ep}}\right) = E_{K/Ep}\left(\frac{\sigma_{c/Ep}}{E_{c/Ep}}\right)$$

When $\sigma_{G/Ep} = 1.29$ Gpa other two layers will have the following stress.

$\sigma_{c/Ep} = 2.32$ GPa

$\sigma_{K/Ep} = 1.55$ Gpa

By the time glass fiber layer reaches the breaking limit other two layers would have already broken.

When $\sigma_{k/Ep} = 1/24$ GPa, other two layers will have the following stress.

$\sigma_{G/Ep} = 1.03$ GPa

$\sigma_C/Ep = 1.86$ Gpa

by the time Kevlar fibre layer reaches the breaking limit the carbon fibre would have broken.

When $\sigma_{car/Ep}$ = 0.97 GN/m^3 the other two layers have the following stress.

$\sigma_{G/Ep}$ = 0.54 GPa

$\sigma_{k/Ep}$ = 0.65 Gpa

This is the limiting case. When carbon fibre layers reaches the breaking limit the other two layers are intact.

Load = 0.97 x 10^3 (0.3 x 0.004) + 1.19 x 10^3 (0.3 x 0.003)

= 2.24 MN

after the carbon fibre layer breaks and becomes ineffective, the Kevlar fibre layer will reach the limit.

Load = 1.24 X 10^3 (0.3 x 0.003) + 1.03 x 10^3 (0.3 x 0.003)

= 2.04 MN

(15) Short carbon fibres with a diameter of 10 mm are to be used to reinforce nylon. If the design stress for the composite is 300 MN/m^2, calculate the load transfer length for the fibres and also the critical length. V_f = 0.3. Interfacial shear strength = 4 MPa.

Material	**E**	**T.S.**
C-Fibre	230 GPa	2.9 Gpa
Nylon 66	2.8 GPa	

(16) Solve problem 15 using the following data.

Material	**E**	**T.S.**
C-Fibre	400 GPa	2.1 Gpa
PEEK	3.8 GPa	0.058 GPa

(17) Calculate the maximum and average fibre stresses for glass fibres which have a diameter of 15 mm and a length of 2.5 mm. The interfacial shear strength is 4 MPa and l_t/l = 0.3.

$$l_t = ((\sigma_f)max\ .\ d)/\ 2\tau_y$$

$$l_c = (\sigma_{fu}\ d)/\ 2\ \tau_y$$

(18)

Material	**ρ (kg/m^3)**	**E(GPa)**
Carbon Fibre	1.9	390
Glass fibre	2.55	72
Epoxy resin	1.15	3
Polyester resin	1.15	3
Steel	7.9	200
Concrete	2.4	45

Find ρ_c and E_c for the following
C_f / E_m with $V_f = 0.6$
G_f / p_m with $V_f = 0.7$
$Steel_f / Concrete_m$ with $V_f = .03$

(19) A composite consisting of 40% by volume uniaxially aligned continuous fibres in a polyester resin supports a stress of 100 MPa parallel to the fibres. If $E_f = 75$ GPa, $E_m = 3.5$ GPa and $\nu_f = 0.21$ and $\nu_m = 0.35$ determine the strains that will occur in longitudinal and transverse directions. What is the modulus of the composite in the transverse direction?

(20) A composite consists of 40% volume of aligned short fibres in a polymer matrix. The diameter of fibre is 7 μm with an aspect ratio of 73. and tensile strength of 4 GPa. The polymer matrix has T.S. of 70 MPa and the interfacial shear strength (τ) is 35 MPa. Calculate the composite tensile strength and the mean stress in the fibre when it fails.

(21) A composite of carbon fibre and nylon matrix has the following data $V_f = 0.2$, $l = 600$ μm, $d = 5$ μm and $(T.S)_f = 3.2$ GPa $(T.S)_m = 65$ MPa and $\tau = 30$ MPa Find the strength of the composite.

(22) Seventy tests on 20 mm long carbon fibre gave a mean strength of 2.1 GPa and a coefficient of variation (Standard deviation divided by mean) of 20%. Estimate the Weibull modulus.

(23) Show that the theoretical maximum volume fraction for identical continuous fibres of circular cross section in a composite material is 0.91. How does this figure relate to volume fraction in a real composite material?

(24) Compare the longitudinal and transverse stiffness of two composites with the same matrix but different fibres if in the first case, $(E_f/E_m) = 50$ and in the second case, $(E_f/E_m) = 25$. $V_f = 0.5$.

(25) Calculate V_{min} and V_{cri} when a metal matrix of T.S = 100 MPa, Y.S = 60 MPa is reinforced with Boron fibre of T.S. = 2.8 GPa.

(26) In a PMC, the matrix is nylon 6.6 and the reinforcement is a continuous carbon fibre. The fibre should carry 95% of the load, applied longitudinally. Using the data given below, find the following,

(a) V_f (b) $T.S_{CL}$ (c) E_{CL}

Material	E	T.S
C_f	2.60 GPa	1.7 GPa
Nylon 6.6	2.8 GPa	76 MPa

(27) An aluminium alloy of Y.S 70 MPa, T.S 100 MPa and Modulus 70 GPa is reinforced with 30% of carbon fibres of strength 2.9 GPa and Modulus of 230 GPa. Estimate $T.S_{cL}$, E_{cL}, E_{cT}, V_{min}, V_{cri} and the fraction of load carried by fibre.

(28) A unidirectional fibre reinforced composite contains discontinuous fibre of length 1 cm and diameter 100mm. Its T.S is 6 GPa. The matrix has a Y.S.=75 MPa and UTS = 110 MPa. What is the strength of the composite if $V_f = 0.5$. What will be the strength if fibre length in reduced to 1 mm? Assume load transfer function to be 0.5.

(29) Determine V_{min} and V_{cri} when a polymer matrix of T.S = 21 MPa and Y.S = 20 MPa is reinforced with aramid fibre of T.S. = 2.8 GPa.

(30) An aluminium alloy matrix of yield strength 80 Mpa, T.S. 125 MPa and modulus 70 GPa is reinforced with Boron fibre, (V_f = 40%) of T.S 2.8 GPa and modulus 380 GPa. Estimate the following:

(a) T.S. of composite in longitudinal direction.
(b) Young's modulus for isostress and isostrain condition.
(c) Fraction of applied load borne by fibres.
(d) V_{min} and V_{cri}.

(31) A fibre reinforced composite has the following properties.
$\sigma_f = 6$ GPa, $\sigma_m = 100$ MPa, $\tau_{my} = 80$ MPa, $d_f = 100$ μm, $V_f = 0.4$
Assuming the load transfer function to be 0.5, determine the strength of the composite when length of the fibre is (a) 5 cm and (b) 5 mm.

(32) A SiC coated Boron fibre (Borsic) is used to reinforce an aluminium matrix. Using the data given below determine the following.
(a) E - parallel to fibre axis (b) E - perpendicular to fibre axis
(c) Strength of composite parallel to fibre.

Material	**E**	**T.S**	**Volume**
Borsic fibre	380 GPa	2.8GPa	30%
Aluminium matrix	70 GPa	35 GPa	70%

(33) A laminate is made of 3 layers. The matrix is epoxy in all three but fibre for the three layers are, alumina, SiC and Boron respectively. Following data is given.

Material	**T.S. GPa**	**E GPa**	V_f	**Dimension (1xbxt) mm^3**
Epoxy	0.07	6		
Alumina	1 4	380	0.5	1000 x 300 x 10
SiC	2.2	250	0.6	1000 x 300 x 10
Boron	3.5	420	0.4	1000 x 300 x 5

Find the load carrying capacity of the laminate when the load is applied on the (b x t) face in the direction of fibre axis.

(34) Solve problem 33 considering Nylon 66 with T.S = 0.07GPa and E=2Gpa in place of Epoxy.

(35) Solve problem 33 considering PEEK matrix with T.S = 0.062 GPa and E = 3.8 GPa in place of Epoxy.

(36) Solve problem 33 considering polyester with T.S = 0.05 GPa and E = 3.0 GPa in place of Epoxy.

(37) Solve problem 33 considering Alluminium alloy with. T.S = 125 MPa and E = 70 GPa as matrix instead of Epoxy.

CHAPTER 14

APPENDIX

1 DIRECTIONAL SOLIDIFICATION OF EUTECTIC ALLOYS

The eutectic reaction ($L \rightarrow \alpha + \beta$) is an invariant reaction and will occur only when each of the three phases involved has a specific composition at a constant temperature. From the phase diagram one can identify the existence of a eutectic reaction, composition of the two solid phases and a liquid phase constituting the reaction and their weight fraction, but one can't predict whether strengthening or reinforcement of one solid phase by the other solid phase will occur on solidification. One of the phases will be continuous (forming the matrix) while the other may be of discrete and discontinuous particles. The discontinuous phase may be of particles shaped like rods, plates dendrites, globs, swords or of intricate, irregular patterns. Regardless of their shape, these are aligned parallel to the direction of growth of eutectic solid from liquid and control the properties. Eutectic solidification if directional, gives a composite in a one step production process. Some examples of such composites are Cu_m / Cr_w (Cr =2%), Al_m /Al_3 Ni (Ni = 10%) and Al_m / $CuAl_2$ (Cu = 50%). The second phase should be an intermetallic or a refractory and should be in the rod form rather than lamellar shapes, for improved properties. A metal and a refractory carbide forming an eutectic solid such as Ta/Ta_2C (C = 8%) or Cb/Cb_2 C (C = 1.5%) also constitute a useful combination. Such directionally solidified eutectic composites show only a gradual decrease in strength as the temperature is increased. No strength deterioration on account of metallurgical instability occurs as the temperature increases.

2 LAMINATES

A composite sheet or tape when cut into various parts, each of same shape and size and arranged one above the other in such a way that the fibre or reinforcement is arranged in the same fashion in every layer and is bonded together, forms a unidirectional laminate. It will be anisotropic. To make the properties less directional, different arrangements can be envisaged. For example, the reinforcement may be perpendicular in alternate layers. It is designated as 0/90 laminate. This way, transverse properties are improved at the expense of axial longitudinal properties. The layers can be arranged in such a way that in the first layer, fibre is in one direction and in the second layer, it is at 90° (say longitudinal) to that in the first layer. The third layer it is at +45° to the first layer, the fourth is at -45° to the first layer, the fifth is, same as the fourth and the sixth is same as the third. Seventh is, again, same as second and eighth is same as the first. If the sequence is repeated the properties of the laminate will be quasi isotropic. It is indicated as 90°/ 0°/ +45°/ -45°/ -45°/ +45°/ 0°/ 90° or simply as o°/ 90°/ ±45°. As the thickness of the laminate increases the strength will asymptotically approach the isotropic in plane strength. One can also have 0°/ 90°/ 0°/ 0°/ 90°/ 0° or 0°/ 90°/ $0°_2$/ 90°/ 0° or $[0 / 90/ 0]_s$ where s stands for symmetrical repeated stacking. $[0°/ \pm 60°]_s$ is also possible. Refer figure 14.1

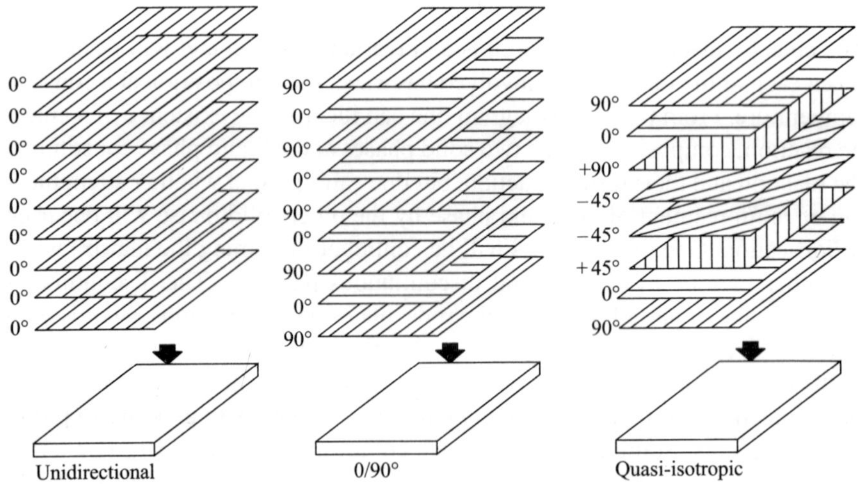

Figure 14.1 The most commonly used lay-ups of UD prepregs

Some of the common laminates are

ARALL	—	Aramid fibre Reinforced Aluminium Laminates
GLARE	—	Glass fibre Reinforced Laminates
CARALL	—	Carbon fibre Reinforced Laminates

3 RULE OF MIXTURES AND PROPERTIES OF COMPOSITES

(1) Poisson's Ratio (v)

$$v_{cL} = v_f V_f + v_m V_m$$

(2) Young's Modulus (E)

$$E_{cL} = E_f V_f + E_m V_m$$

$$\frac{1}{E_{cT}} = \frac{V_m}{E_m} + \frac{V_f}{E_f}$$

(3) Thermal Expansion Coefficient (α)

$$\alpha_{cL} = \frac{\alpha_m E_m V_m + \alpha_f E_f V_f}{E_m V_m + E_f V_f} \text{ (for fiber composite)}$$

$$\alpha_{cT} = (1 + \nu_m)\, \alpha_m V_m + (1 + \nu_f)\, \alpha_f V_f - \alpha_{cL}\, \nu_{cL}$$

$$\alpha_{cL} = \frac{\alpha_m V_m K_m + \alpha_p V_p K_p}{V_p K_p + V_m K_m} \text{ (for particulate composite)}$$

(4) Thermal conductivity (λ)

$$\lambda_{cL} = \lambda_m V_m + \lambda_{f_1} V_f$$

$$\lambda_{cT} = \frac{\lambda_{f_2} \lambda_m}{\left(\lambda_{f_2} V_f + \lambda_m V_m\right)}$$

λ_{f1} = – for fibre in axial direction

λ_{f2} – for fibre in transverse direction

(5) Density (ρ)

$$\rho_c = \rho_f V_f + \rho_m V_m$$

(6) Dielectric Constant (ω)

in series

$$\frac{1}{\omega_{cs}} = \frac{V_f}{\omega_f} + \frac{V_m}{\omega_m}$$

in parallel

$$\omega_{cp} = V_f \omega_f + V_m \omega_m$$

4 LIST OF SOME COMPOSITES

Polymer Matrix Composites

$Epoxy_m$ — $Aramid_f$

$Epoxy_m$ — $Carbon_f$

$Polyetherimide_m$ — $Carbon_f$

$Epoxy_m$ — $Glass_f$

$Polyetherimide_m$ — $Glass_f$

$Epoxy_m$ — $Graphite_f$

$Epoxy_m$ — Al_2O_{3f}

$Epoxy_m$ — $Glass_f$

$Nylon66_m$ — $Graphite_f$

$Nylon66_m$ — $Carbon_f$

$Nylon66_m$ — $Glass_f$

$PTFE_m$ — $Carbon_f$

$PTFE_m$ — $Glass_f$

$Polysulfone_m$ — $Glass_f$

$Polysulfone_m$	— $Carbon_f$
$Polyimide_m$	— $Graphite_f$
$Epoxy_m$	— SiC_f
$Epoxy_m$	— B_f

Metal Matrix Composites

Al_m	— Al_2O_{3f}
Al_m	— B_f
Al_m	— SiC_f
Mg_m	— Al_2O_{3f}
Ti_m	— SiC_f
Cu_b	— SiC_f
Al_m	— G_f
Mg_m	— G_f
Cu_m	— W_f
Al_m	— W_f
$St.St_m$	— W_f
Al_m	— SiC_p
6061 Al_m	— SiC_p
2124 Al_m	— SiC_p
7090 Al_m	— SiC_p
7091 Al_m	— SiC_p
Al_m	— SiC_w
2124 Al_m T6	— SiC_w

Ceramic Matrix Composites

Al_2O_{3m}	— SiC_w
Al_2O_{3m}	— SiC
Si_3N_{4m}	— SiC_f
$Glass_m$	— SiC_f
Glass $Ceramic_m$	— SiC_f
Glass $Ceramic_m$	— Al_2O_{3f}
ZrO_{2m}	— SiC_w
$MoSi_{2m}$	— SiC_w
$Mullite_m$	— SiC_w
Si_3N_{4m}	— SiC_w
SiC_m	— TiC_p
Si_3N_{4m}	— TiC_p
Al_2O_{3m}	— TiC_p
Si_3N_{4m}	— ZrO_{2p}
Al_2O_{3m}	— ZrO_{2p}
Al_2O_{3m}	— SiC_p
$Glass_m$	— C_f
$Glass_m$	— Al_2O_{3f}
Al_2O_{3m}	— Al_2O_{3f}
SiC_m	— SiC_f

5 INTERFACIAL BONDING IN COMPOSITES

Matrix wets reinforcement and is in intimate contact with it. Bonding takes place between the two. The type of bonding varies from system to system. More than one type of bonding can exist. Mechanism of bonding may change during various stages of service or production. There are broadly four types of bonding.

(1) MECHANICAL BONDING

Interlocking or keying of the matrix surface and reinforcement surface takes place.

(2) ELECTROSTATIC BONDING

One surface is positive and the other negative, electrically. As a result of attractive electrostatic forces between such oppositely charged surfaces, bonding takes place. This mechanism is effective over very short distances (of the order of atomic distances).

(3) CHEMICAL BONDING

This bond is due to attraction between two chemical species (here matrix and reinforcement). To help in coupling, between matrix and reinforcement another species may be added. For example coupling between a polymer matrix and a glass fibre can be enhanced by a silane.

(4) REACTION OR INTER DIFFUSION BONDING

Atoms or molecules of matrix will interdiffuse into those of reinforcement and vice versa at the interface. Diffusion is important at high temperatures and over longer time. The extent of inter diffusion depends on temperature and time. In this case the bonding force will be the highest and the bond will be the strongest.

6 APPLICATIONS OF COMPOSITES

TYPE	APPLICATIONS
MMCs	Diesel engine piston, Microelectronic housing, Racing bicycle frames, Golf club shafts, Structural tubes for space applications, Neutron shielding material, Heat dissipator Cold plate material, Portable bridge components, Wing structural elements, Small lightweight pressure vessel applications, Fins for projectiles, Missile body casings, Drive shafts for engines, Tubes, Turbine discs, Hollow fan blades, Propellers that generate low noise, Wires, Antennae, Thin walled tubes, Strips, Rods, Angles, Channels, Plates, Hollow cylinders, Flywheels, Simply loaded beams, Turbine blades, Pressure vessels, Turbine vanes. Precision components, Gyroscope parts, Infrared mirror, laser mirrors, Instrument covers, Missile wings, Floor panels in a test bed aircraft, Horse shoes for race horses, Connecting rod, Brake discs and drums, Scroll compressor, Struts, Antennae, Superconductors, Contacts, Filaments, Electrodes.
PMCs	Gear case for railway locomotives, Doors and door flames, Axial flow fans, Sleepers for railway girder bridges, Pultruded profiles such as gratings, solid rods for electrical insulation, Cable trays, Ladders, Orthopaedic applications, Shutters for windows & toilets, Axle box covers

(Contd.)

(*Contd.*)

TYPE	APPLICATIONS
	in railway coaches, Boards for partitions, Paneling, Roofing, Cupboards & doors, Sleeper berth backing, Coachbody, Luggage bin, Seating. Leaf springs, Gears, Bearings, Body panels, Pultruded drive shafts, Springs for heavy trucks & trailers, Fuel tanks, Structural panel covering torsion box, Battery trays, Truck boxes, Pickup Boxes, Helmets, Two wheeler body parts, Minesweeper hulls, Sheet processing rolls, Helicopter rotor blades, golf driving club, Tennis racquets, Pole vault poles, Aircraft structural parts, Fibre glass fishing boats, Skis, Archery bows, Surf boards, Fishing Rods, Canoes, Diving Boards, Playground equipment, Wings, Fuselage, Radomes, Landing gears, Nose cones, Ablative shields.
CMCs	Cutting tools, Turbine blades, Tube Vanes, Seals, Bearings, Supersonic planes, Space planes, Wear guides, Automotive engine components, Nozzles, Exhaust ducts, Heat exchanger tubes, Recuperators, Combustors, Centrifuges, Filters, Substrates, Air preheaters, Wire drawing & extrusion dies, Valve seats, High precision balls, Bearings, Plungers for chemical pumps, Liquid propellant rocket motors, Turbojet parts, Thermal structures for space planes & other space probes, Gas turbine combustor cans, Brakes, After burners, Heatshields, Prostheses, Fixation plates, Thermal insulation, Rocket nozzles.

7 LIST OF SYMBOLS

$P(x)$ — Probability density function
$p(x)$ — Cumulative distribution function
S — Weibull reliability function
M — Weibull modulus
X_u — Location parameter
X_o — Scale parameter
V — SpecimenVolume
V_o — Reference Volume
L_α — Length of α layer
Δl_α — Increase in length of α layer
L_β — Length of β layer
Δl_β — Increase in length of β layer
ε_α — Strain in α layer
ε_β — Strain in β layer
ε_c — Strain in composite
ε_{fu} — Strain in fibre corresponding to σ_{fu}
ε_{mu} — Strain in matrix corresponding to σ_{mu}
ε — Strain
ε_0 — Instantaneous strain
ε_∞ — Strain at infinite time
N — Total number of layer
Δl_c — Increase in length of composite

L — Breadth and width
F — Force applied
σ_{α} — Stress in α layer
σ_{β} — Stress in β layer
σ_c — Stress in composite
σ_m — Stress in matrix
$\sigma_{m\varepsilon c}$ — Stress in matrix when strain in composite is ε
$\sigma_{m\varepsilon f}$ — Stress in matrix when strain in fibre is ε
σ_{fr} — Fracture strength
σ_{max} — Maximum stress
$\sigma_{f\varepsilon c}$ — Stress in fibre when strain in composite is ε
$\sigma_{f\varepsilon f}$ — Stress in fibre when strain in fibre is ε
$\sigma_{c\varepsilon c}$ — Stress in composite when strain in composite is ε
σ_f — Average stress in fibre
σ_{fu} — Ultimate strength of fibre
σ_{mu} — Ultimate strength of matrix
σ_{fy} — Yield strength of fibre
σ_{my} — Yield strength of matrix
T.S. — Tensile strength
τ_m — Shear stress in matrix
τ_{my} — Shear yield stress of matrix
d_f — Diameter of fibre
l_c — Critical length of fibre
l — Length of fibre
P — Load
S — Span length
B — Breath
D — Depth
V_m — Volume of matrix
V_f — Volume of fibre
V_{α} — Volume of α layer
V_{β} — Volume of β layer
V_c — Critical volume fraction
V_{α} — Volume fraction of α layer
V_{β} — Volume fraction of β layer
V_{min} — Minimum Volume fraction
E_{α} — Young's Modulus of α
E_{β} — Young's Modulus of β
E_c — Young's Modulus of composite
τ_I — Interfacial shear strength
σ_I — Interfacial tensile strength
R — Radius of fibre
E_f — Young's modulus of fibre
E_m — Young's modulus of matrix

I	—	Final intensity of ray after absorption
I_o	—	Initial intensity of ray
X	—	Thickness of composite material
μ	—	Linear absorption co efficient of material
Z	—	Acoustic impedance of material
ρ	—	Density
C	—	Sound wave velocity in material
I_T	—	Transmitted intensity of ultrasonic wave
I_R	—	Reflected intensity of ultrasonic wave
w_m	—	Weight of matrix
w_f	—	Weight of fibre
W_m	—	Weight fraction of matrix
W_f	—	Weight fraction of fibre
ε_{sm}	—	Steady state creep rate of matrix
ε_{sc}	—	Steady state creep rate of composite
ε_{sf}	—	Steady state creep rate of fibre
ε	—	Creep rate
ECT	=	$0.5\ T_m$ = Equicohesive temperature
T_m	—	Melting point in Kelvin
(l/d)	—	Aspect ratio of fibre
G	—	Shear Modulus
G	—	Crack extension force
G_C	—	Critical crack extension force
ΔQ	—	Activation Energy
k	—	Boltzman Constant
T	—	Temperature
A	—	Constant
N	—	Constant
N_f	—	Number of cycles in fatigue
S	—	Stress in fatigue
σ_α	—	Stress amplitude
σ_r	—	Range of stress
σ_m	—	Mean stress
$\sigma_{\max}$	—	Maximum stress
$\sigma_{\min}$	—	Minimum stress
$\left(\dfrac{da}{dN}\right)$	—	Increase in crack length per cycle
A	—	Constant
p	—	Paris exponent
ΔK	—	Stress intensity factor
ν	—	Poisson's ratio
ρ	—	Density
E_L	—	Longitudinal Young's Modulus

E_T	—	Transverse Young's Modulus
α	—	Thermal expansion coefficient
λ	—	Thermal conductivity
C	—	Heat capacity
Ω	—	Resistivity
$T.S_{cL}$	—	Longitudinal tensile strength of composite
E_{cL}	—	Longitudinal Young's modulus of composite
B	—	Breadth of specimen in fracture toughness testing
W	—	Width of specimen in fracture toughness testing
ω_m	—	Dielectric constant of matrix
ω_f	—	Dielectric constant of fibre
ω_{cs}	—	Dielectric constant of composite (in series)
ω_{cp}	—	Dielectric constant of composite (in parallel)
K_{Ic}	—	Fracture Toughness
a	—	Crack length
γ_s	—	Surface energy
γ_p	—	Energy for plastic deformation

ARALL	—	Aramid fibre Reinforced Aluminium Laminate
GLARE	—	Glass fibre Reinforced Laminate
CARALL	—	Carbon fibre Reinforced Aluminium Laminate
DMC	—	Dough Moulding Compound
SMC	—	Sheet Moulding Compound
CSM	—	Chopped Strand Mat
CFRP	—	Carbon Fibre Reinforced Plastic
GFRP	—	Glass Fibre Reinforced Plastic
KFRP	—	Kevlar Fibre Reinforced Plastic
G_f	—	Glass fibre
Gr_f	—	Graphite fibre
LAS	—	Lithium Alumino Silicate
PAN	—	Poly Acrylo Nitrile
YAG	—	Yittrium Aluminium Garnet
FP	—	Alumine fibre

QUESTIONS

CHAPTER 1

1. Define a composite material.
2. Classify composite materials on the basis of (i) matrix & (ii) reinforcement.
3. What are the benefits of composites?
4. What is a (i) Laminate (ii) Lamina?
5. What is a hybrid composite?
6. Why PMCs enjoy a preeminent position in the field of composites?

CHAPTER 2

1. What are the metallic fibers that are commonly used? How are they produced? Give their important & relevant properties?
2. How are Kevlar Fiber and Polyethylene Fuser manufactured?
3. How are SiC (i) Fiber (ii) Whisker & (iii) Particle are processed?
4. Give the equation for production of boron fiber. Draw a sketch of the process & explain.
5. What is the method for producing glass fibers
6. Give the various routes of making alumina fiber.
7. Define a composite fiber.
8. How carbon fiber is manufactured from (i) PAN (ii) PITCH as starting materials? Discuss?
9. Compare critically the properties of various fibers.

CHAPTER 3

1. What are the two classification of routes for producing MMCs?
2. What are the disadvantages of P/M processing of MMCs?
3. What are the advantages and disadvantages of diffusion bonding?
4. Why is melt stirring not a preferred route for MMCs?
5. How is melt stirring improved?
6. Compare squeeze casting and liquid melt infiltration by gas pressure.
7. What is the time taken for spray co-deposition?
8. Insitu process doesn't give a true composite in the strict sense of definition of a composite. Why?
9. Why lighter metals are more often chosen as matrix materials?
10. Name a few applications of MMCs.
11. Name a few MMCs.
12. Name a few metals used as matrices in MMCs.

CHAPTER 4

1. Why PMCs are preeminent compared to MMCs and CMCs?
2. Name the three classes of polymer matrices.
3. Compare the similarities and dissimilarities of hand and moulding methods.
4. What are disadvantages of hand methods?
5. Name the advantages of moulding methods.
6. Draw the sketches for bag moulding methods and explain.
7. Explain RTM.
8. What is the advantage of pultrusion method?
9. What are dry winding and wet winding?
10. Name a few applications of PMCs.
11. Name a few PMCs.
12. What is fiber glass?

CHAPTER 5

1. Why CMCs are difficult to process by conventional melting and casting route?
2. Name a few ceramics used as matrices for CMCs.
3. Name a few conventional and new techniques of producing CMCs.
4. What are single action pressing and double action pressing in P/M route of processing CMCs?
5. Why hot pressing is suited for glasses and glass ceramics materials?
6. Explain reaction bonding? Which composite is produced by this method
7. Explain matrix transfer moulding with a sketch.
8. What is the disadvantage of pultrusion method of producing a CMC?
9. Analyse the principle of Lanxide process.
10. Name the two advantages of CVI.
11. What is the temperature required for CVI of (i)Si_3N_4 (ii) Al_2O_2 matrices?
12. Define (a) Sol (b) Gel.
13. What are the staring materials for Sol-Gel processing of (a) SiC matrix (b) Si_3N_4 matrix?
14. Give the advantages of polymer pyrolysis.
15. Explain (a) Polymer pyrolysis (b) SHS.
16. Name a few applications of CMCs.

CHAPTER 6

1. Name the aerospace applications of CCCs.
2. Why a coating is required for CCCs to be used at high temperatures?
3. Describe CVD of CCCs.
4. Explain pulsed CVD.
5. Discuss pyrolysis of (i) Thermo sets (ii) Thermo plastics.

6. Name a few thermal insulation applications of CCCs.
7. Name a few recreational applications of CCCs.

CHAPTER 7

1. Name a few intermetallic alluminides.
2. Name a few intermetallics other than aluminides.
3. What is an ordered intermetallic?
4. What are the properties of IMCs?
5. Discuss the production of IMCs?
6. How toughness of IMCs can be improved?
7. Give the futuristic applications of IMCs.

CHAPTER 8

1. Derive the VFR or ROM for continuous fiber composites in isostress & isostrain conditions.
2. Derive an expression for V_{min} & V_{cvi} for the continuous fiber composites.
3. What is critical length of discontinuous fiber?
4. Define the critical aspect ratio of discontinuous fiber.
5. Derive an expression for (l_c/d_f).
6. Derive VFR or ROM for discontinuous fiber composites when $l>l_c$ & $l<l_c$.
7. Discuss the stress v/s strain curve for CMCs PMCs & MMCs .
8. Draw the stress vs strain curve for following matrices .
   Alumina, (Ceramic) Aluminium (metal) PEEK (Polymer thermoplastic) Epoxy (Polymer thermo set) and Rubber.
9. Draw the the stress vs strain curve for following fibers Boron, SiC, Al_2O_3, Kevlar, glass, and carbon.
10. Draw schematic stress vs strain curve for CMCs, PMCs, MMCs, and CCCs.

CHAPTER 9

1. Explain what you under stand by ECT.
2. Discuss the circumstances under which creep can occur.
3. The steady state creep rate of a composite follows ROM. True or False.
4. Show creep strain Vs time for CMCs, PMCs & MMCs.
5. CMCs have better creep resistance. State True or False. Why?
6. What are the 3 types of cyclic loading that can cause fatigue?
7. Show the S-N curve for CMCs, PMCs, MMCs, and CCCs. What is Parislaw?
8. Is Paris law obeyed by composites?
9. Elaborate on fatigue of CMCs, PMCs, MMCs, and CCCs.
10. Elucidate on fatigue of (i) Laminates (ii) Hybrids.
11. What are the three modes of fracture?

12. How K_{IC} is related to crack length 'a'?
13. How is K_{IC} determined in lab?
14. CMCs have poor $K_{IC.}$ Why?
15. MMCs have good K_{IC} Why?
16. Name the possible toughening mechanisms that are operative composites.

CHAPTER 10

1. How are the following carried out on composites (i) tensile test (ii) compression test (iii) flexural test & (iv) shear test?
2. What is the significance of each of the above test?
3. What are the types of interfacial shear strength test? Explain each one of them.
4. How is micro indentation test carried out? Explain.
5. Impact test measures toughness of a material. True or false. What is the importance of toughness?
6. Creep test is relevant at high temperature. Define high temperature for a given metal or material.
7. Fatigue test is important when there is cyclic loading at low stress. True or false.
8. What is the principle of radio graphy? State in three or four words.
9. What is the principle of ultra sonics? State in three or four words.
10. How NDT results are corroborated?
11. What are sensitnity, resolution, accuracy and reliability in NDT?
12. Name the benefits of NDT to industry and society?
13. Who developed weibull's statistics?
14. What are (i) Shape parameter (ii) Scale parameter (iii) Location parameter?
15. Define (i) Probability density function & (ii) Cumulative distribution function.
16. What is the significance of Weibull modulus? Explain with sketches.
17. Define survival probability.
18. Define error in testing.

CHAPTER 11

1. Define a dispersion strengthened composite (DSC).
2. Explain the processing techniques used for production of dispersion strengthened composites.
3. Name a few DSCs.
4. Name a few applications of DCs.
5. Enumerate the properties of DCs.
6. Define a particulate strengthened composite (PSC).
7. Explain the processing techniques used for production of particulate strengthened composites.
8. Name a few PSCs.

9. Name a few applications of PSCs.
10. Enumerate the properties of PSCs.
11. What is a cermet?
12. What are the advantages of cermets?
13. What are the applications of cermets?
14. Describe the production methods for cermets.
15. Name a few cermets.
16. What is a polymet?
17. Name a few applications of polymet.
18. How do you produce a polymet?

SOLVED NUMERICAL EXAMPLES

An aluminium alloy matrix of yield strength of 70 MPa, tensile strength of 125 MPa and modulus of 70 GPA in reinforced with 50 vot% of boron filament of tenside strength 2.8 GPa and modulus of 380 GPa estimate (i) tensile strength of composite in longitudinal (ii) the modulus in longitudinal and transverse direction (iii) V_{min} & V_{cri} & (iv) fraction of applied load taken up by fiber.

Yield stress of matrix σ_{ym}	–	70 MPa
Ultimate stress of matrix σ_{ym}	=	125 MPa
Young's Moduless of Matrix E_m	=	70 GPa
Terminal stress of Fiber σ_{uf}	=	2.8 GPa = σ_{ym}
Youngs Modulus of fiber E_f	=	380 GPa
Volume Fixetion fibre V_f	=	0.5
Tensile strength of composite σ_{cL}	=	$\sigma_f V_f + \sigma_m V_m = 2800 \times 0.5 + 1.25 \times 0.5 = 1462$ MPa
Longitudinal Yongs Modules of composite E_{cL}	=	$E_f V_f + E_m V_m = 380 \times .5 + 70 \times .5 = 225$ GPa

$$\frac{1}{E_{CT}} = (V_f / E_f) + (V_m / E_m)$$

$$= \left(\frac{0.5}{70} + \frac{0.5}{380} \right)$$

Transverse Young's Modules of composit $E_{CT} = 118.22$ GPa

$$\text{Critical vilum fraction of fiber } V_{\text{cri}} = \frac{(TS)_m \; \sigma_{m\varepsilon f}}{(TS)_f + (TS)_m - \sigma_{m\varepsilon f}}$$

$$= \frac{125 - 70}{2.8 \times 10^3 + 125 - 70} = 0.0192$$

Minimum on volume fraction of fiber

$$V_{\text{mini}} = \frac{(TS)_m \; \sigma_{m\varepsilon f}}{(TS)_f - \sigma_{m\varepsilon f}}$$

$$= \frac{125 - 70}{2.8 \times 10^3 - 70} = 0.02$$

Load on fiber $F_f = \sigma_f A_f$

$$= E_f \varepsilon_f V_f V_c$$

$$F_{CL} = E_f \varepsilon_f V_f V_c + E_m \varepsilon_m V_m V_c$$

$$= \frac{F_1}{F_{CL}} = \frac{E_f V_f}{E_f V_f + E_m V_m} = \frac{380.5}{70 \times .5\ 380 \times .5} = 0.84$$

(2) Compare the longitudinal and transverse stiffness of 2 composites with same matrix and different fibers when $V_f = 0.5$ and modulus ratio is 50 for one composite and 25 for another

Longitudinal Modulus

$$E_{CL} = E_f V_f + E_m V_m$$

$$= \frac{E_{CL}}{E_m} = \frac{E_f}{E_m} V_f + V_m$$

Composite I = $\frac{E_{cLL}}{E_m} = 50 \times 0.5 + 0.5$

= 25.5

Modulus of composite I $E_{cI} = 25.5\ E_m$

Composit II $\frac{E_{cII}}{E_m}$ $25 \times .5 + 0.5$

= 13

Modulus of composite II

$E_{cII} = 13\ E_m$

$$\frac{E_{cI}}{E_{cII}}$$

Transverse Modulus

$$= \frac{1}{E_{CL}} = \frac{V_f}{E_f} + \frac{V_m}{E_m}$$

$$= \frac{E_m}{E_{CL}} = \frac{E_m}{E_f} = V_f + V_m = \frac{1}{50} \times .5 + .5$$

Modulus of composite I $E_{CTI} = 1.96\ E_m$

$$= \frac{E_m}{E_{CTI}} = \frac{E_m}{E_{fII}} = V_f + V_m = \frac{1}{25} \times 0.5 + 0.5$$

Modulus of composite II $E_{CTII} = 1.92\ E_m$

$$= \frac{E_{CTI}}{E_{CTII}} = 1.02$$

(3) Density of composite made from unidirectional glass fiber in an epoxymatrix in 1950 Kg/m^3. If density of glass fiber is 2.540 Kg/m^3 and density of epoxy is 1300 Kg/m^3 calculate the volume fraction of fiber and weight fraction of fiber.

Density of composite $\rho_c = \rho_f V_f + \rho_m V_m$

$1950 = 2540\ V_f + 1300\ (1 - V_f)$
Volume fraction $(V_f) = 0.524$

$$\text{Weight fraction} = 68.25\% = \left(\frac{V_f \cdot \rho_f}{\rho_c}\right)$$

(4) A unidirational fiber reinforced composite is made with discontinuous fiber of length 10mm dia = 100 μm and strength of 6 GPa. Matrix has a yield strength of 80 MPa and tensile strength of 100 MPa. What is the strength of composite if $V_f = 0.5$. What will be the strength of composite if fiber length is reduced to 1 mm

$$\tau_y = \sigma_y / 2$$

$$\text{Crtical length of fiber } l_c = \frac{\sigma_{fu}\ df}{2\tau_y} = \frac{6 \times 10^3 \times 10 \times 10^{-6}}{2 \times 40}$$

$l_c = 7.5$ mm

$l = 10$ mm

$l > l_c$

Tensile strength σ_c of composite $= V_f + \sigma_{m\varepsilon c}\ V_m$

$$= 6 \times 10^3 \left[1 - \frac{7.5}{2 \times 10}\right] 0.5 + 100 \times 0.5$$

$\sigma_c = 6 \times 10^3\ [1 - .375]\ 0.5 + 100 \times 0.5$

$= 6 \times 10^3\ .625 \times 0.5 + 100 \times 0.5$

$= 6 \times 10^3 \times .3 + 100 \times .5$

$= 1850 + 50$

$= 1850$ MPa

$l < l_c$ when $l = 1$ mm

Tensile strength of composite

$$\sigma_c = \sigma_{fec} \left(\frac{1}{21c}\right) V_f + V_m\ \sigma_{m\varepsilon c}$$

$$= 6 \times 10^3\ \frac{1}{2 \times 7.5}$$

$= 250$ MPa

BIBLIOGRAPHY

1. Engineered Materials Handbook Vol.1 ASM International, 1987.
2. L. J. Broutman & R. H. Krock, Modern Composite Materials Addison & Wesley, Reading, 1967.
3. E. D. Rawlings & F. L. Mathews, Composite Materials – Science & Engineering, Chapman & Hall, London, 1994.
4. K. K. Chawla, Ceramic Matrix Composites, Chapman & Hall, London, 1993.
5. K. K. Chawla, Composite Materials, Sringer Verlag, Newyork, 1987.
6. Minouru Taya & R. J. Arsenault, Metal Matrix Composites, Pergammon Press, Oxford, 1989.
7. Bryan Harris, Engineering Composite Materials, IoM London, 1999, Wood Head 2006.
8. R. J. Crawford, Plastic Engineering, Pergammon Maxwell Macmillan International, 1987.
9. H. Buhl, Advanced Aerospace Materials Springer – Verlag, Berlin 1992.
10. N. S. Stoluff & V. K. Sikka, Physical Metallurgy & Processing of Intermetallic Compounds, CBS, India 1997.
11. D. Hull & T. W. Clyne, An Introduction to Composite Materials, Cambridge University Press, 1996.
12. Peter C. Powell, Engineering with Fibre, Polymer Laminates, Chapman & Hall, London, 1994.
13. G. Savage, Carbon – Carbon Composites, Chapman & Hall, London, 1993.
14. T. W. Clyne & P. J. Withers, An Introduction to Metal Matrix Composites, Cambridge University Press, Cambridge, 1995.
15. I. M. Ward & D. W. Hadley, Mechanical Properties of Solid Polymers, John Wiley & Sons, Chichester, 1993.
16. Richard Warren, Ceramic Matrix Composites Blackie Glasgow, 1992.
17. Handbook of Reinforcements for Plastics, John V. Milewski & Harry S. Katz, Van Nostrand, Newyork, 1987.
18. Navin Chand & M Fahim, Tribology of FRP Materials, Allied Publishers, New Delhi, 2000.
19. Michel Barsoum, Fundamentals of Ceramics, McGraw Hill, New York, 1997.
20. Thomas H. Courtney, Mechanical Behaviour of Materials, McGraw Hill, 1990.
21. G.E.Dieter, Mechanical Metallurgy, McGraw Hill, London, 1988.
22. Proceedings of International Conference on "Advances in Composites – 2000, 24-26 Aug., 2000, Bangalore, India.

23. P. Ramarao, Advances in Materials and their Applications, Wiley Eastern, New Delhi, 1993.
24. A.K.Kaw, Mechanics of Composite Materials, CRC Press, New York, 1997.
25. K.S.Mazdoyasni, Fibre Reinforced Ceramic Composites, Noyes Publishers, 1990.
26. Anthony Kelly, Composite Materials Concise Encyclopedia, Pergammon Press, Oxford, 1889.
27. Mel Schwartz, Composite Materials Hand book, McGraw Hill, 1992.
28. A.G.Bratukhin & V.S.Bogolyubar, Composite Manufacturing Technology, Chapman & Hall, London, 1995.
29. Geoff Eckold, Design & Manufacture of Composite Structures, Woodhead Publishers, Cambridge, 1994.
30. S.R.Finn & G.S.Springer, Composite Plates Impact Damage, Technomic, Lancaster, 1991.
31. P. Ramakrishnan, Advances in Composite Materials, Oxford & IBH, New Delhi, 1991.
32. Plastics & Rubber Institute, Automated Composites, Sci. & Tech Publishers, 1986.
33. Leonard Hollaway, Handbook of Polymer Composites for Engineers, Woodhead Publishers Ltd., Cambridge, England, 1994.
34. DST India, Polymers & Composites, Oxford & IBH, New Delhi, 1989.
35. J.W.Weeton, D.M.Peters & K.L.Thomas, Engineers Guide to Composite Materials, ASM, Ohio, 1987.
36. J.Delmonte, Metal/Polymer Composites, Van Nostrand Reinhold, New York, 1990.
37. D.V.Rosato, D.P.Dimattic & D.V. Roasanto, Designing with Plastics & Composites, Van Nostrand Reinhold New York, 1991.
38. T.Richarson, Composites A Design guide, Press Inc., 1987.
39. Mel Schwartz Emerging Engineering Materials, Technomic, Lancaster, 1996.

INDEX